D0765512

Dino Gangs

DR PHILIP J CURRIE'S NEW SCIENCE OF DINOSAURS

Dino Gangs

By Josh Young

All images courtesy of Atlantic Productions except for the following:

p20 Friedrich Saurer/Science Photo Library; p31 James Steinberg/Science Photo Library; p34 Joe Tucciarone/Science Photo Library; p37 Ria Novosti/Science Photo Library; pp48, 97, 157 Dr Philip J Currie; pp52–3 Steve Gschmeissner/Science Photo Library; pp62–3 Jaime Chirinos/Science Photo Library; p91 bottom American Museum of Natural History; p114–15 Philippe Psaila/Science Photo Library; p130 Julius T Csotonyi/Science Photo Library; pp139, 166 bottom Discovery FootageSource; p143 bottom Silver Phoenix LLC/Footage Search; p205 Absolutely Wild Visuals; p241 BBC Motion Gallery; pp288–9 Jose Antonio Pe As/Science Photo Library.

Computer-generated imagery pp143 top, 293 top and bottom, 300, 301, 302 created by ZOO (www.zoovfx.com).

First published in 2011 by Collins
HarperCollins Publishers
77–85 Fulham Palace Road
London W6 8JB

www.harpercollins.co.uk

13 12 11
8 7 6 5 4 3 2 1

A catalogue record for this book is
available from the British Library

ISBN: 978-0-00-741339-3

Printed and bound in Great Britain by
Clays Ltd, St Ives plc

This book is dedicated to those children whose eyes grow as big as saucers when they see their first dinosaur exhibit, and then follow that passion into a career in dinosaur palaeontology.

contents

introduction

the terrible lizards

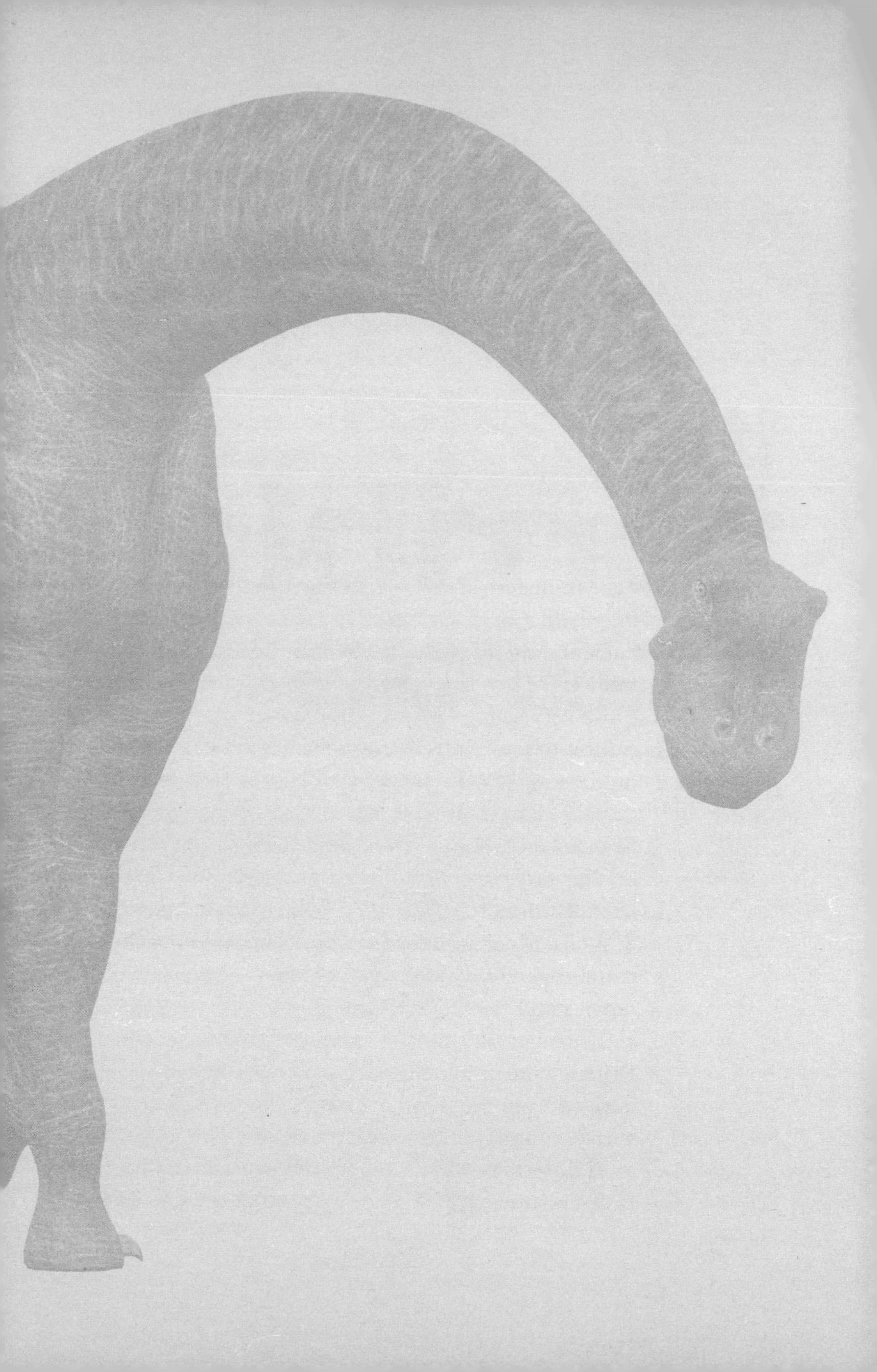

s the monsters of our nightmares, dinosaurs have long held a unique place in our fascination with ancient animals. Even though dinosaurs lived millions of years ago, they somehow seem very real – and very scary. Anyone who has seen a museum exhibit of a mounted dinosaur skeleton cannot help but be awed by the sheer size of some dinosaurs and the bizarre features of others. Dinosaurs have been used to market everything from sweets and breakfast cereal to petrol and pasta, and they have been featured in movies and books as demonic dragons. Much of what we know about dinosaurs comes from how they are portrayed in books, movies and popular culture.

The term 'dinosaur' was coined in 1842 by the British palaeontologist Richard Owen. It was derived from the Greek words *deinos*, meaning 'terrible', and *sauros*, meaning 'lizard'. Not long after dinosaurs were scientifically identified, Charles Dickens introduced a *Megalosaurus* in the opening paragraph of his 1853 novel *Bleak House* as something one would not want to see walking up the street. In his popular 1912 novel *The Lost World*, Sir Arthur Conan Doyle, creator of Sherlock Holmes, set out a scenario under which dinosaurs ruled a tropical paradise in South America. Edgar Rice Burroughs used them repeatedly in his 'lost world' stories and featured them on the cover of his 1924 novel *The Land That Time Forgot*. The cinema brought dinosaurs to life in the classic monster movies *King Kong* and *Godzilla*. Indeed Godzilla was created to look like a large dinosaur, with its massive head and lower body resembling that of a *Tyrannosaurus* and its dorsal plates modelled after those found on a *Stegosaurus*.

Phil Currie studying a Tyrannosaurus *skull and its large teeth used for crushing bone.*

Novelist Michael Crichton, the best-selling author of *The Andromeda Strain* and *Congo*, wrote a thriller in 1990 about what would happen if dinosaur DNA that had been sucked out by mosquitoes and preserved for millions of years in amber was used to create dinosaurs that could live in the modern day. Using an entertaining blend of science and fiction, *Jurassic Park* showed that every dinosaur for itself meant that humans were in deep trouble. The mega bestseller was made into a film in 1993 by Steven Spielberg, the most commercially successful filmmaker of all time with hits such as *Jaws* and *E.T.: The Extra-Terrestrial*.

Jurassic Park created a worldwide dinosaur mania. The film became the highest grossing movie ever at the time, earning more than $1.2 billion worldwide at the box office. The film's 1997 sequel, *The Lost World: Jurassic Park*, was the second highest grossing movie worldwide that year, and the movie-going public was terrorized yet again by the modern-day dinosaurs in 2001 with *Jurassic Park III*, another huge box-office hit. Universal Studios also built an elaborate *Jurassic Park* ride, complete with a 15-metre (50-foot) *Tyrannosaurus rex*, at its Hollywood theme park and a *Jurassic Park* River Adventure at its Orlando amusement park, which have attracted millions of visitors.

As accurate as parts of *Jurassic Park* were, such as promoting dinosaurs as the ancestors of birds, the filmmakers also took some artistic licence to up the ante. For starters, the novel's and film's premise is based on a real scientific improbability. The film also had some of its dinosaurs living in the wrong time period. *Tyrannosaurus rex*, nicknamed *T. rex*, and *Velociraptor*, known by its catchier nickname 'Raptor', were the featured dinosaurs in the movie, and yet both lived millions of years later in the Cretaceous period, rather than the earlier Jurassic period. And the message of the movie – that if you had a fast truck you could escape the dinosaurs because they were slow – appears not to be true.

The *Jurassic Park* phenomenon was a bellwether event that shined the spotlight on dinosaurs, and by extension on the work done by devoted palaeontologists who spend their lives trying to piece together how the ancient beasts lived and died more than 65 million years ago. Museums with dinosaur exhibits saw a 40 per cent spike in attendance in 1993, and there was little drop-off in the ensuing years.

Noted palaeontologist Phil Currie embraces popular culture as a way of generating public interest in dinosaurs. In fact, through a friend Currie was able to enlist Crichton to write the introduction to his academic *Encyclopedia of Dinosaurs*.

'I love science, but I love science fiction, too,' Currie says. 'It was pretty uncanny what they did in a lot of ways, both Crichton originally and Spielberg subsequently. Crichton was very good at creating suspense and hooking you. I'm sure he had an awful lot of fun writing these books because he got to delve into something to the degree that very few people ever do except for the scientists or sociologists.'

Currie has been infatuated with dinosaurs since he was six years old, and his obsession with these creatures remains to this day. He is one of the world's most highly regarded palaeontologists, the Indiana Jones of the dinosaur trade. In fact, Currie even sounds like Harrison Ford, the actor who played Indiana Jones. Tall and slender with tousled grey hair, Currie lives like a man on a mission to find out more about the beasts he loves. He is so afraid of his focus being diverted that he doesn't own a mobile phone or accept voicemail messages. He saves times by walking up – and down – the stairs two at a time, and to ensure not a moment is wasted, every appointment must be run through his wife and assistant, Eva Koppelhus, a palaeontologist in her own right.

Currie has impeccable credentials: he has served as a museum curator and one of the founders of the Royal Tyrrell Museum, one of the premier dinosaur museums in the world, and he is a distinguished professor holding a Canada Research Chair at the University of Alberta in Edmonton, a city literally built on dinosaur remains. Between 1986 and 1990 he served as co-director of the Canada–China Dinosaur Project, the first palaeontological partnering (in the Gobi Desert) between China and the West since the Central Asiatic Expeditions of the 1920s. He has helped describe some of the first feathered dinosaurs, named many carnivorous dinosaurs and found some of the first dinosaur eggs in North America. He is one of the primary editors of the influential *Encyclopedia of Dinosaurs*, and has written groundbreaking scientific papers and non-fiction books as well as a series of fictional children's books on dinosaurs with his wife. He also teamed up with Microsoft's Nathan Myhrvold to construct a computer model that showed that sauropods ('lizard-footed' large plant eaters) actually broke the sound barrier with their mighty tails as they swung them to scare off their enemies. But for all his stature in the palaeontology world, Currie keeps a practical, child-like approach to his trade, because to him, dinosaurs are 'really, really cool'.

For the past 15 years, Currie has been on a physical and intellectual journey to piece together a jigsaw puzzle that could change the way we think about massive killer dinosaurs for ever. He believes that tyrannosaurids, the most fearsome family of carnivorous dinosaurs, were far more complex and more dangerous than we ever could have imagined.

'Part of the problem is just that we always tend to lump dinosaurs together,' Currie says. 'We know dinosaurs existed for 150 million years and we think that they didn't go through any changes. But *Tyrannosaurus* was the top predator of the very end of the age of dinosaurs. Those were the most highly adapted, large predators that existed during the entire age. And I think they were by far the most dangerous animals.'

The tryannosaurid family had a dozen species. They were part of the Coelurosauria clan, which contained all bipedal, theropod ('beast-footed') dinosaurs. Most were small meat-eaters, such as oviraptorids, ornithomimids and dromaeosaurids (including *Velociraptor*), and some dated back to the Jurassic period. Tyrannosaurids came from animals similar to *Velociraptor* and were relatively late in the timeline of dinosaur existence.

Currie has focused on a member of the tyrannosaurids that hasn't drawn much attention: *Tarbosaurus*, which lived in the Gobi Desert, was a cousin of the fierce *Tyrannosaurus rex*, an inhabitant of North America. Imagine a beast 12 metres (40 feet) long, 3–4 metres (10–13 feet) tall and weighing in at 5–6 tonnes. Its massive skull was packed with

'Tyrannosaurs were the most highly adapted large predators and ... by far the most dangerous animals.'

64 giant, serrated, bone-crushing teeth, and it also had what appears to be one of the biggest brains of any of the large meat-eating dinosaurs. That was *Tarbosaurus*.

'*Tarbosaurus* is probably best known because of its cousin, *Tyrannosaurus rex*, and for a long time after *Tarbosaurus* was discovered, it was actually called *Tyrannosaurus* because it looks very similar in most ways,' Currie explains, referring to the dinosaurs in the present tense as he often does. 'So we have a large animal that kind of looks like an over-grown chicken on hormones. If this thing is standing on its two hind legs, basically it can't use its forelimbs at all for locomotion. In the case of *Tyrannosaurus*, they become very reduced and small although they still have fairly big claws and muscles. *Tarbosaurus* is even crazier in having even smaller arms.'

Tarbosaurus was discovered in 1948 and was originally named *Tyrannosaurus bataar* (and later *Tarbosaurus bataar*) by a Russian scientist named Evgeny Maleev in 1953. A few scientists still consider these the same animals, but Currie points out that there are fundamental differences. 'They are animals that look very much the same because they are mostly the same size, but they are not the same,' he says. 'The easiest difference to pick up on is the arms of the animals. *Tyrannosaurus* has really short arms, but *Tarbosaurus* has even more ridiculously short arms.'

Next to *Tyrannosaurus rex*, *Tarbosaurus* is the largest tyrannosaur that existed, which makes it one of the top three or four giant predators ever in the world. *Tarbosaurus* was an animal with a very large skull, and this skull was very powerfully muscled. Over time, the back of the skull expanded laterally so that it could build up muscle mass to allow it to close its massive jaws in a powerful bite. Its teeth were the size of bananas, and the reason those teeth became so thick is so they could function as bone crushers, rather than steak knives. The teeth were able to slice through flesh because the jaws had so much power in them and because they were sharp tipped. At the same time, the thickness allowed them to break bones, which they swallowed with the flesh. This was a dinosaur that was very well adapted for eating other animals.

Tarbosaurus had very powerfully built hind legs that were relatively long. 'When you look at not just the length of the legs but the proportions of the legs, tarbosaurs are built more like ostriches than human beings,' Currie says. 'This meant that they were capable of reaching very high speeds for their size.'

Tarbosaurs had also adapted their bodies so that they weren't quite as heavy as one would expect them to be. Rather than resembling a small meat-eating dinosaur that had been blown up to a larger size, its body proportions had changed considerably to prevent it from having too much weight to haul around. That excess weight would have been a problem because these animals ran on their hind legs, making them like seesaws.

'This was a really nasty animal, and in spite of its size and that it looked so primitive to us, this was also a dinosaur that was very, very sophisticated in a lot of ways. If you look at the brain size of this animal, it's not as big as a human being, but if you compare the brain size of a *Tarbosaurus* to the brain sizes of any of the animals it was chasing down, then it had a pretty respectable brain size. Even compared to modern lizards and crocodilians, turtles and things like that, this dinosaur had a fairly big brain. The bottom line was that it was a pretty sophisticated dinosaur and a very efficient killing machine.'

In 2009, Currie went on a voyage of discovery, revisiting places and ideas from his 30 years in

A Tyrannosaurus *looked like an overgrown chicken on hormones.*

palaeontology to reveal one of the last major unknowns in dinosaur biology. He was accompanied to dinosaur sites and high-tech labs by several internationally renowned scientists: Dave Eberth, a Canadian sedimentologist and habitat specialist; Yoshi Kobayashi, an associate professor at Hokkaido University who is Japan's top expert in carnivorous dinosaurs; Yuong-Nam Lee, a Korean herbivorous dinosaurs specialist; Louis Jacobs, an American college professor and renown vertebrae fossil expert; and Larry Witmer, an American scientist and professor engaged in cutting-edge research into dinosaur brains.

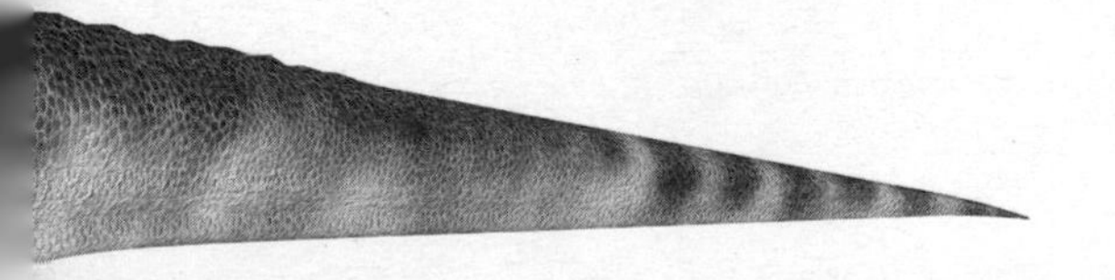

In addition to debating and fleshing out his theories with the experts, Currie wants his findings to be presented to the widest possible audience. He seeks to dispel the notion that studying dinosaurs is conducted by scientists holed up in windowless labs with microscopes examining ancient bones that have little relevance. Instead, he hopes to show that dinosaurs were not just fossilized stones but living, breathing creatures that may have quite a bit to say about the history of our planet.

chapter 1

the world of dinosaurs

hat exactly is a dinosaur? Dinosaurs were a diverse group of animals that first appeared some 225 million years ago and lived on all seven continents, with more than 1000 species having been identified. Though they had a slight resemblance to lizards, dinosaurs distinguished themselves from other animals on the planet at the time because many of them could walk on two legs. The primary characteristics that scientists use to classify these terrestrial animals were an upper leg bone with a ball and socket joint at the hip, a streamlined lower leg bone and an ankle that functioned as part of the lower leg bone.

Dinosaurs roamed and dominated the planet for more than 160 million years during the Mesozoic era, making them one of the most successful animal types ever.

The Mesozoic era is divided into three periods: the Triassic (225 million to 213 million years ago), the Jurassic (213 million to 144 million years ago) and the Cretaceous (144 million to 65 million years ago). Dinosaurs were at their most diverse and evolved in the Late Cretaceous period, the time when tyrannosaurids ruled the Northern Hemisphere, and it is this period that is the primary focus of many palaeontologists.

At the end of the Cretaceous, dinosaurs had been around for some 160 million years and were at the peak of diversity. The different groups were highly specialized in many ways. They lived everywhere on the planet, all the way from the Arctic to the Antarctic at that time, in environments that were quite diverse. In North America, there were coastal regions and areas that were closer to the mountains that were rising at the time, and there were also dry areas and deserts in some places. Each one of those habitats was home to different groups of dinosaurs because each dinosaur species was adapted for specific climates and environments.

Size is often the first thing people think of when dinosaurs are mentioned. Though the dinosaurs portrayed in popular culture are usually shown to be huge, the fact is that most dinosaurs were the size of humans or even smaller. Because the fossil record is incomplete, scientists sometimes use educated guesswork to estimate the absolute size of the biggest and smallest dinosaurs.

The longest dinosaur is believed to have been *Seismosaurus*, which lived during the Late Jurassic period. *Seismosaurus* was a sauropod, or plant-eater, with a long neck and swooping tail that from head to tail measured between 35 and 40 metres (115–135 feet) long. The length is estimated because only part of the best *Seismosaurus* skeleton was recovered from its northern New Mexico site and the specimen is still being prepared. Despite *Seismosaurus*' impressive length, it weighed less than 30 tonnes – heavy but nowhere near the heaviest sauropods (found in Argentina) that are believed to have weighed as much as 100 tonnes. (The largest carnivores were *Tyrannosaurus rex* and *Giganotosaurus carolini*, both of which lived in the Cretaceous period, the former in North America and the latter in South America. They each measured about 13.5 metres (45 feet) long and weighed about 6 tonnes.) By comparison, a blue whale can grow to 30 metres (100 feet) in length and weigh a shocking 180 tonnes.

On the opposite end of the scale, a herbivore find that was aptly named *Mussaurus*, or 'mouse lizard', would have fit in the palm of an adult's hand. *Compsognathus*, or 'elegant' dinosaur, one of the smallest adult carnivores, was the size of a large chicken and weighed approximately 3 kilos (7 pounds). *Microraptor* was an even smaller carnivore find in China from the Early Cretaceous.

Dinosaurs fall into two basic categories: herbivores (plant-eaters) and carnivores (meat-eaters). The plant-eaters greatly outnumbered the meat-eaters. Some of the best-known herbivores were all quadrupeds, including sauropods, anklyosaurs, stegosaurs, hadrosaurs and horned dinosaurs like *Centrosaurus*. Sauropods had long necks, tiny heads and massive bodies. Ankylosaurs were very wide-bodied with massive plates of bone covering them like armour. Hadrosaurs are commonly called duckbilled dinosaurs because their mouth looked very much like a modern duck's bill. *Stegosaurus* was distinguished by rows of bones along its back that developed into plates. *Centrosaurus* is one of the most common ceratopsians, or horned dinosaurs, which can be

'Fossils – objects that have gone through permineralization.'

identified by their unique skull features not found elsewhere in the animal kingdom. On the tip of the upper jaw is a rostral bone, which forms what looks like a parrot's beak.

The best-known bipedal carnivores are *Allosaurus*, *Velociraptor*, *Albertosaurus* and *Tyrannosaurus*. *Allosaurus*, or 'different lizard', was a large predator in the Jurassic period with extremely sharp teeth and it often measured over 8.5 metres (28 feet) long. *Velociraptor* and its closest relatives (the 'Raptors') were feathered dinosaurs, most of which only grew to the size of large dogs. Each had four claws on each foot, one of which was adapted as a can opener and used to disembowel prey. *Albertosaurus* is noteworthy for its two-fingered hands and massive head containing dozens of large, sharp teeth. Despite being a top predator in its area and weighing more than 2 tonnes, *Albertosaurus* didn't come close to measuring up to the monstrous *Tyrannosaurus*.

Dinosaurs were given their names by the scientists who described them in scientific papers. Because all languages are different, the names are then translated into Latin or ancient Greek, the common languages of scientists the world over, despite being 'dead' tongues. There is no set way to name a dinosaur. Some of the names focus on a characteristic of the dinosaur and others on how it might have lived. For example *Tyrannosaurus rex* means 'tyrant lizard king', while *Tarbosaurus* translates as 'terrifying lizard'. Other names refer to the locations where they were discovered; *Albertosaurus* was first found in Alberta, Canada. Dinosaurs have also been named as tributes to people, such as *Othnielia*, which was named in honour of palaeontologist Othniel C. Marsh.

What scientists know about dinosaurs has come from fossils – objects that have gone through permineralization, the process by which minerals are deposited in the pores of bones and turn to stone. Fossilization is a very fickle process. The sediments and groundwater must be right for preservation. There must be an accumulation of sediments and no rain and wind washing them away over the millions of years. In the mountains, erosion prevents long-term accumulation of fossils, because if bones were buried, it wouldn't be for very long because the sediments would be washed out of the mountains into the low lands.

Fossil is derived from the Latin word meaning 'dug up', and this is truly the case for dinosaur fossils. Most evidence of dinosaurs comes from original bones infilled with minerals, rather than from imprints of them frozen in time or bone that has been replaced by stone. In some cases, dinosaur bones were encased in ironstone nodules after they were buried, and this protected the bones from water-carrying minerals in solution so that to this day the fossils look just like modern animal bones. Others lines of evidence come from footprints, eggs and even skins that have been fossilized and preserved in stone. Scientists are able to date many dinosaur fossils from the rocks they are found in, and this has enabled them to establish an accurate timeline for dinosaurs despite the fact that they lived millions and millions of years ago.

The dinosaur timeline was established largely through a process called radiometric dating of fossils. This involves comparing radioactive isotopes to the decayed material found in the same rocks or the surrounding layers of rocks. In North America, significant volcanic activity created ash beds that contain radioactive material. In this case, when scientists find a layer of ash, then a layer of fossils, then another layer of ash, they know that the fossils between those two ash beds are bracketed by the two dates, the lower bed providing the older date and the higher bed the younger. They can then

RIGHT
Dinosaur footprints. Four trackways of dinosaurs moving to the upper left, and at least one other dinosaur coming from the upper right and one from the lower left.

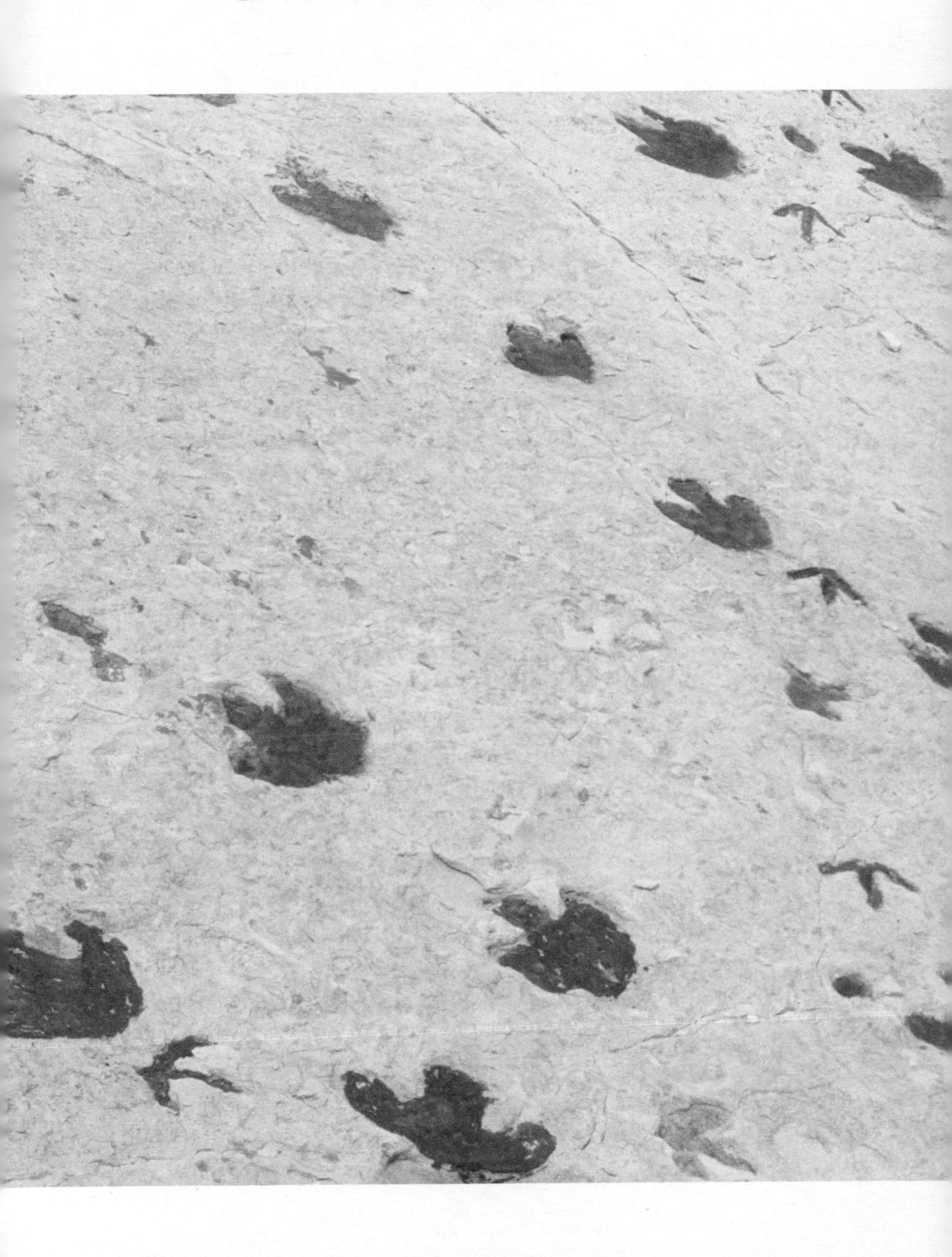

compare those fossils to similar ones found in Europe and conclude the Europeans ones are the same age. The radiometric dating techniques used provide dates of plus or minus 10 per cent accuracy, so 90 million years would have left a 9-million-year margin of error. But more sophisticated testing now provides dates to plus or minus a couple of hundred thousand years. In human age, that's not very close, but in terms of dinosaurs, it is.

Fossilized dinosaur footprints have been helpful to palaeontologists in determining what kinds of dinosaurs lived in certain areas. Though they can only rarely identify the species of the animal through footprints, palaeontologists can tell the general type of dinosaur that made the prints. Footprints are particularly revealing in situations where there are consecutive footprints that continue in one direction. These are called trackways, and they have enabled palaeontologists to draw both physical and behavioural conclusions about dinosaurs. Trackways reveal that most dinosaurs walked upright and did not drag their tails. They also show evidence of which dinosaurs were living together. And in certain situations, the stride lengths in the trackways can be measured and used to estimate speed. Without fossilization, this would not be possible.

* * *

There are long-simmering controversies among palaeontologists about the details of how dinosaurs were born, how they lived and how they died. Although there is no exact way to determine how long each species of dinosaur lived, scientists estimate that most species lasted between 2 and 5 million years. Their life spans varied by species and size. Some of the larger dinosaurs, such as *Allosaurus*, are believed to have lived for around 50 or 60 years, while smaller ones like *Compsognathus* may have lived for only 5 to 10 years. *Velociraptor* is estimated to have lived for about 20 years, and *Tyrannosaurus* and *Tarbosaurus* about 30.

One of the most hotly debated questions was whether or not dinosaurs were warm-blooded. The advances in that area changed dinosaur science and offered a prelude to more complex theories.

In the late 1960s, palaeontologist John H. Ostrom led the way in arguing that dinosaurs were warm-blooded. Ostrom was a professor at Yale University and in his later years served as Curator Emeritus of Vertebrate Paleontology at the Peabody Museum of Natural History. This was a radical idea at that time. Ostrom's 1964 discovery and subsequent study of *Deinonychus* led him to conclude that the animal's horizontal posture and sleek body, combined with the sickle-shaped claw on each foot, dubbed the 'terrible claw', offered convincing evidence that it was an active predator with a high metabolism. He could also see that it looked very much like *Archaeopteryx*, the first bird. He had also found multiple specimens of *Deinonychus* in the same quarry in Montana. This caused him to assert that small, meat-eating dinosaurs were behaviourally complex and may have lived in packs. Ostrom's student, Robert Bakker, further argued these characteristics meant that dinosaurs were, in fact, warm-blooded.

These revolutionary theories changed the way dinosaurs were shown and started what Bakker later dubbed the 'dinosaur renaissance', a period of study that eventually would double our recorded

Velociraptor.

knowledge of dinosaurs. 'All of those ideas were coming out of that one find,' Phil Currie says.

There are now many lines of argument that dinosaurs were warm-blooded. One is bone histology, or microscopic anatomy. 'The structure of dinosaur bones is very much like the structure of mammal or bird bones, and it's quite different than what we see in reptiles,' Currie says. However, he cautions that, by itself, bone histology doesn't prove warm-bloodedness, because it may just indicate very active growth. 'Then you argue you don't have active growth unless you can sustain it, and you don't sustain it unless you have warm-bloodedness,' he adds.

The predator–prey ratio is another line of evidence. Dinosaur finds indicate a 5 per cent predator-to-prey ratio. This indicates that the predators were very active and had to maintain a low ratio of predators to prey otherwise they would consume all the available food. The opposite is seen in cold-blooded animals, where the predator-to-prey ratio can be as high as 50 per cent. Cold-blooded animals like snakes can survive a month on one

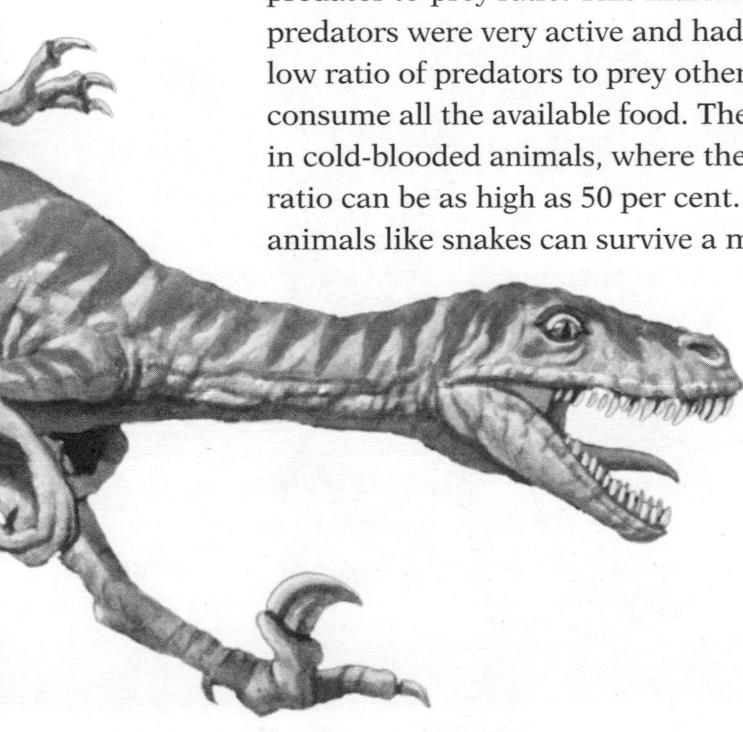

meal, meaning that their environment can sustain far more predators.

Dinosaur-egg finds, coupled with nesting habits, have added another layer of evidence to the argument that dinosaurs were warm-blooded, and they have also helped palaeontologists study parental behaviour.

The very first dinosaur eggs reported scientifically were found in southern France and England back in the 1800s. However, at the time they were believed to be bird eggs, and therefore they were not properly categorized, despite the fact that they were discovered with dinosaur bones. The first identified dinosaur eggs were discovered by explorer Roy Chapman Andrews in Mongolia in 1923. At that point, scientists re-examined the eggs found in France and England and determined they were dinosaur eggs as well.

The early finds of dinosaur eggs were confused with bird eggs because the knowledge base and number of finds were both limited. Now scientists can easily tell a bird egg from a dinosaur egg by viewing a slice of the egg under a microscope and studying its anatomy. 'If you look at an eggshell closely, you can see that there are pores going through it and in fact there is a crystalline structure to it,' Currie explains. 'Every species of animal has a different crystalline structure to its eggshell. Birds have eggs that are very close to dinosaur eggs, but they are a little bit different in their layering. Basically, dinosaur eggs have an extra layer in the

Preparing a nest of fossilized dinosaur eggs, originally discovered in Mongolia.

crystalline structure – dinosaur eggs have two layers and bird eggs have a third layer on the outside, which generally isn't found in most dinosaur eggs.'

Scientists have looked to dinosaur eggs for clues not only about how they reproduced but how they lived. Noted palaeontologist Jack Horner, who has published numerous scientific papers on dinosaurs and served as technical advisor to the movie *Jurassic Park*, and his colleague Bob Makela found some of the first evidence that dinosaurs cared for their young. Horner and Makela studied nests of dinosaur eggs discovered in the 1970s in Montana and dated to the Late Cretaceous. The large quantity of nests and the spacing between the nests led them to conclude that these dinosaurs nurtured their babies by protecting the nests and bringing them food. Horner and Makela named the new species of dinosaur *Maiasaura*, which means 'caring mother lizards'. In 1983 in the same area, which was nicknamed 'Egg Mountain', they found similar egg sites for the carnivorous dinosaur *Troodon*.

'What we learned from Egg Mountain, since we had *Maiasaura*, which are duckbill dinosaurs, and *Troodon*, which are carnivorous, was that these dinosaurs were doing basically the same thing,' Horner says. 'While there was no evidence that *Troodon* kept babies in their nests, they certainly kept them on their nesting horizon. These nesting grounds suggest that both groups of dinosaurs were social. We haven't really had evidence for any dinosaurs of any kind since that would suggest they weren't social. Basically, every group of dinosaur has been found in accumulations together, suggesting that they lived in some kind of aggregation.'

Another egg breakthrough occurred in 1987 when Currie and his team found the first hadrosaur eggs containing embryos in southern Alberta. 'Identifying dinosaur eggs has only [come about] in the last 30 years because before that there was no association between eggs, embryos and adults,' Currie explains.

The examination of eggs has also heightened the public's interest in dinosaurs by making them seem more accessible. Of course, part of the fascination with dinosaurs is that they are big so many people are surprised that dinosaur eggs are relatively small. Still, the realization that dinosaurs had babies is compelling.

'For a long time, people had just thought of dinosaurs as being some kind of rocks, not so much as being animals that lived or breathed,' Currie says. 'When you know they have babies and are very much like modern animals in so many ways. Even within the eggs – we could see that the babies had been moving their jaws so they could grind down their teeth – it says something about the biology of the animals, which brings it home big time.'

The study of the eggs has revealed how some dinosaurs nested and protected their eggs. Oviraptorids, which were small, roughly human-sized dinosaurs, were originally known as 'egg snatchers'. In the 1920s, the plant-eating *Protoceratops* was the most common kind of dinosaur at the site where the eggs were found, so the conclusion was that the nest of eggs found with the oviraptorid skeletons belonged to *Protoceratops*. But as palaeontologists made more oviraptorid egg finds, they realized that the dinosaurs were, in fact, protecting their own eggs, not stealing those of other dinosaurs.

'We now know that in the case of oviraptorids that they stayed in one spot when laying eggs. First, they laid a pair of eggs, then they turned a little and

'Egg finds offer compelling evidence that dinosaurs were warm-blooded.'

laid another pair next to the first pair. And then they turned to lay another pair, and so on. So basically they stood on one spot and laid a circle of eggs around their feet,' Currie explains.

He stands in the middle of his office, expands his arms like wings, and excitedly acts out the scene. 'At the same time they are doing that, they took their hands and scooped sand onto the eggs. Once the first layer was down, they lay a second layer and then sometimes a third layer, all of which were buried in sand. This process created a trench around the nest,' he continues, mimicking how the ancient animal might have done it. 'In some cases, we know precisely how they were laying their eggs.

In the case of oviraptorids, it was in 1990 when we found our first nest with a mother on top; since then at least four more nests have been found with the mothers sitting on top.'

Currie believes that in addition to egg finds showing how the mother protected her eggs, they offer compelling evidence that dinosaurs were warm-blooded. 'The study of eggs can even tell you something about physiology. If the mother is lying on the nest – birds lie on their nests to keep the eggs at a constant temperature – it may well tell us that these dinosaurs were warm-blooded and in fact were brooding their eggs and keeping their eggs warm.'

The evidence that supports this theory comes from the layout of the eggs in the nest. The eggs formed a circle around the outside of the nest like a doughnut. The mother was standing in the middle, which allowed her chest to cover the eggs in the front and her tail to cover those in the rear. With her arms outstretched around the eggs to the sides, she could also protect those eggs.

'If you look at the feathered dinosaurs where the feathers are behind the arms, the feathers would cover those eggs on top,' Currie says. 'Large dinosaurs couldn't do that – they didn't have large enough wingspans. In the case of these dinosaurs that have been found brooding, it has been suggested that there were several reasons that these long feathers developed, including as a mechanism for shading the nest, for protecting the eggs and/or for keeping them warm. Brooding doesn't work if you are cold-blooded.'

What have not been found are *Tarbosaurus* eggs – yet. 'It's probably because of the fact that they would have been in an environment with acidic groundwater so the eggs were destroyed,' Currie says. 'Of course, there are other possible explanations, one being that they would have nested farther inland in a place where things weren't getting fossilized. Still, we may eventually find a *Tarbosaurus* egg in the right place, where it did have a chance to fossilize. If we are lucky, there will be an embryo inside and we'll be able to put the whole story together.'

Because of their nesting habits and feathers, it has been established that oviraptorids and dromaesaurids like *Velociraptor* were probably warm-blooded. That has led Currie and several other palaeontologists to argue that it therefore makes sense that the closely related *Tyrannosaurus* would have been warm-blooded, though this is still considered somewhat forward thinking.

Currie lays out the argument. '*Velociraptor* and oviraptorids were warm-blooded in all likelihood,' he says. 'They are feathered dinosaurs; it doesn't make any sense to have feathers on your body as insulation unless you are warm-blooded. The advantage [cold-blooded] lizards have is that when they get cold, they just move into the sun and they warm up pretty quick. But if you put feathers on them, it would be like taking an ice cube and wrapping a blanket around it and sticking it outside; it doesn't work. So there are a lot of reasons to think that the little guys are warm-blooded; notably these include the fact that they are so close to the ancestry of birds, and that they have bone histology like modern mammals and birds.'

Tyrannosaurus, like *Tarbosaurus* and *Albertosaurus*, was relatively closely related to *Velociraptor* and oviraptorids, so Currie concludes that because of the fact that they are closely related to these warm-blooded creatures, they were almost certainly warm-blooded, too. 'It doesn't make sense that you have guys in your ancestry who are warm blooded, that all your close relatives are warm-blooded and that your descendants are warm-blooded, but you are not,' he says. 'So it makes sense that tyrannosaurs were warm-blooded.'

As radical as it was to accept that dinosaurs were warm-blooded and laid eggs, it was even more revolutionary to see them as big, non-flying birds rather than just scary, oversized lizards. In 1974, John Ostrom, the palaeontologist who found *Deinonychus*, revised his description after the discovery of a more complete specimen and championed the idea that birds were descended from dinosaurs. Robert Bakker backed it up with additional research, and their work provided the dinosaur renaissance with another major development.

Ostrom's scientific paper reiterated the idea that had been presented a hundred years earlier by British scientist Thomas Henry Huxley, a defender of Darwin's theory of evolution, who had proposed that birds were descended from dinosaurs. Huxley came to this conclusion after studying *Archaeopteryx*, the oldest known fossilized bird. *Archaeopteryx* lived during the Late Jurassic period, and its features suggested that it was a transitional fossil between dinosaurs and birds. Since *Archaeopteryx* was discovered in 1861, only 10 specimens have been found, despite Herculean efforts to find more.

Additional evidence for this theory came in 1986 when Jacques Gauthier, a scientist who is now based at Yale University, published a list of more than 125 characteristics shared uniquely by birds and dinosaurs. Currie calls this very, very powerful evidence under any kind of modern scientific analysis. 'We don't even have such strong evidence for other transitions such as reptiles into mammals,' he points out. 'A lot of palaeontologists by the mid-1980s already believed that birds came from dinosaurs. I got into researching it because of the fact that our Late Cretaceous dinosaurs from Alberta were very bird-like in a lot of ways, and it started me thinking about it and publishing on the subject.'

The theory was slow to make its way into the mainstream. Currie believes this is because most ornithologists don't work on fossils, and only work on modern birds. 'Maybe 50 per cent of the scientists working in palaeontology believed birds came from dinosaurs, maybe 5 per cent of ornithologists believed it, and very few people in the public believed it because they had never heard of it.'

In 1996, Currie himself came face-to-face with the evidence from one of the very first specimens showing the transition in what turned out to be a complicated situation where cultures, science and publishing collided.

The story began in 1994 when a farmer in north-eastern China found a fossil of what was thought to be a species related to *Archaeopteryx* and sold it to a local museum. The farmer made some money on the transaction, so he went out and dug some more and found another fossil, which he sold to a different Chinese museum. Scientists were aware of those finds and figured there were more where those had come from. In 1996, at the Tucson Rock and Mineral Show, there were hundreds of specimens resembling this bird, complete with feathers. Excitement began to grow because this was a bird that had never been scientifically described, and suddenly there were already far more specimens of fossil birds than had showed up in more than 100 years. But were they genuine?

For a long time palaeontologists had talked about the possibility of feathered dinosaurs. This discussion stemmed from the theory that if any dinosaurs were warm-blooded, then the most likely would be small dinosaurs; and if they had any kind of insulation on their bodies, it would probably be feathers. By the late 1970s, many scientists had concluded that dinosaurs gave rise to birds. Though it was mostly speculation and there was scant evidence to prove it at the time, dinosaurs were being illustrated with feathers on their bodies in books everywhere.

Currie first saw a picture of the Chinese's farmer's find in a Beijing newspaper in 1996 after a trip to Mongolia. The article reported that a feathered dinosaur had been found, and the story was accompanied by a picture. 'It was a little picture that didn't look like anything,' Currie recalls. 'The idea of feathered dinosaurs had been with me for a long time, but still, what are your chances?'

Through a contact, Currie made arrangements to see the 'feathered dinosaur' specimen. He was slightly suspicious because he had been given an exact time for his viewing. When he arrived 10 minutes early, his contact walked him up and down the street so they wouldn't be early. At the appointed time, the contact ushered Currie into a room full of reporters. 'I realized that I had walked into a press conference and that they were going to show me this specimen in front of the Chinese press to see what my reactions were.'

Several different specimens were bought out one at a time. The boxes were then opened in front of Currie. 'I would see a beautiful insect with spectacular preservation. Then they would bring out another box, and I would open it up and see a spectacular lizard fossil and so on.'

This dragged on for so long that Currie began thinking it was a diversion and that they weren't even going to show him the feathered dinosaur. 'I went to see it not because I thought it was feathered, but because I could see from the photograph that it was a complete specimen of a small dinosaur,' he says. 'Small dinosaurs are rare; carnivorous dinosaurs are rare. This was obviously a very, very important specimen' – provided it actually existed.

Finally, the box with the feathered dinosaur arrived unannounced. 'When they opened the lid of the box, my eyes probably expanded 20 times like a cartoon character,' Currie remembers. 'First of all,

'My eyes probably expanded 20 times like a cartoon character.'

the specimen was beautifully preserved, but secondly, my eyes were drawn to these things that were around the outside of the body that were supposed to be feathers. In my mind, I had rationalized that it was probably dendrites or some kind of fungal growth. I just didn't think the chances of finding a feathered dinosaur were all that good. Sure enough, within milliseconds, I knew that what I was looking at was real, and in fact, we did have the first feathered dinosaur.'

Fossilized tarbosaur skin.

Whether or not it was a legally collected, genuine specimen would take years to resolve. A week after Currie returned to Canada, he began fielding calls first from Japanese reporters and then from British reporters. Days later, on 19 October 1996, the story hit the front page of the *New York Times* under the headline 'FEATHERY FOSSIL HINTS DINOSAUR–BIRD LINK'. The story was accompanied by a drawing done by Michael Skrepnick, the artist who was travelling with Currie in China, and it reported Currie's assessment that this was, in fact, a feathered dinosaur. 'The whole world went a little crazy for a while,' Currie says.

For years, dinosaur feathers continued to provide something of a mystery for scientists. Currie believes there is a possibility all the meat-eating dinosaurs known as coelurosaurs ('hollow-tailed lizards') had feathers as babies to provide insulation. The big species then shed those feathers as they grew into adulthood and no longer needed the feathers. The larger a dinosaur became the less its surface area was in relation to its mass or volume. Big animals have a problem ridding themselves of excess body heat. However, a small animal would lose heat really fast. So if small animals are warm-blooded, they have to be insulated in some way, such as with feathers or fur. However, a whale or an elephant is so large that it doesn't need the insulation. Could it have been the same for *Tarbosaurus*?

'It may well be that *Tarbosaurus* is free of feathers only because it's big,' Currie says. 'When

Tarbosaurus were born, they were probably only a half metre or 18 inches long. At that stage, they may have needed feathers. So there was a prediction, which is kind of cool, that if we ever find a small *Tyrannosaurus* then it should have feathers because it is closely related to these feathered dinosaurs.'

The cool part, Currie says, is that in 2004 a small *Tyrannosaurus* was found in north-eastern China that was about the size of a German shepherd and it had feathers. Currie explains that the big problem is that there are very few places in the world where conditions are such that feathers would be preserved, though skin impressions are often found in Mongolia and Alberta. Feathers rot away pretty quickly, so typically they decompose before they have a chance to fossilize.

The environment controls what is preserved. In most dinosaur-fossil sites there are no eggs or feathers found. However, in an environment like north-eastern China, where there was a lot of volcanic action, things preserve far better. Volcanic ash would rain down on the lakes. Sometimes the ash would kill a bird or a dinosaur running along the shore, and they would fall into the lake. Because the ash is very fine grained, mixed with the mud in the lake it preserves details very well. More importantly, it alters the chemical environment and kills the bacteria that would otherwise decompose the keratin – the horny material that forms fingernails, beaks and feathers – and leaves these structures preserved for science. 'Suddenly, you have this amazing situation where you not only get fingernails and beaks but also feathers preserved,' Currie says. That kind of preservation is critical for scientists to formulate theories about dinosaur feathers that connect them to birds.

Fossilized feathers have also provided scientists of the first evidence of dinosaur colours. Melanosomes, the biological structures that give feathers colour, were recently found to have been preserved in the small feathered theropod dinosaur seen by Currie in 1996 and subsequently named *Sinosauropteryx*, which lived 125 million years ago. The melanosomes allowed scientists to determine that the dinosaur had a red Mohican with a red and white striped tail. However, scientists have not been able to determine the colour for most other dinosaurs, even those whose skin has been preserved. Early artistic renderings of dinosaurs were in browns and greys and were based on the colour of the larger modern animals such as elephants and Komodo dragons, but scientists still have little to no evidence that dinosaurs were similar in colour to these.

OVERLEAF
A close up of Komodo dragon skin – one of the largest reptiles alive on Earth at present – shows what dinosaur skin may have looked like.

'We are absolutely nowhere with the colour of *Tarbosaurus*,' Currie admits. 'So far all we have is tarbosaur skin, but we don't have any evidence of colour banding to show us that there might be melanosomes preserved. Of course, that doesn't mean that they are not, because what might happen eventually is that somebody might take a look at the

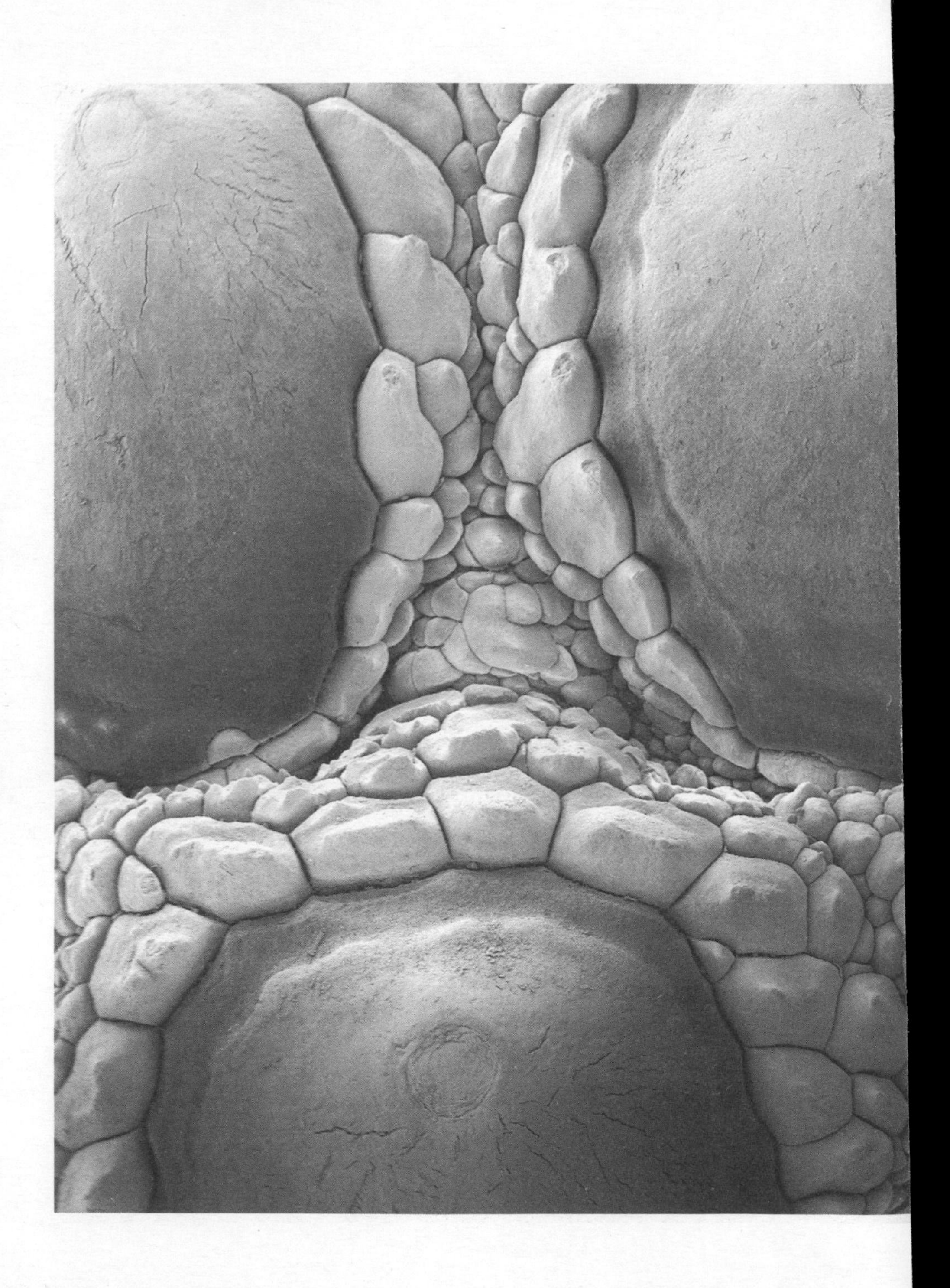

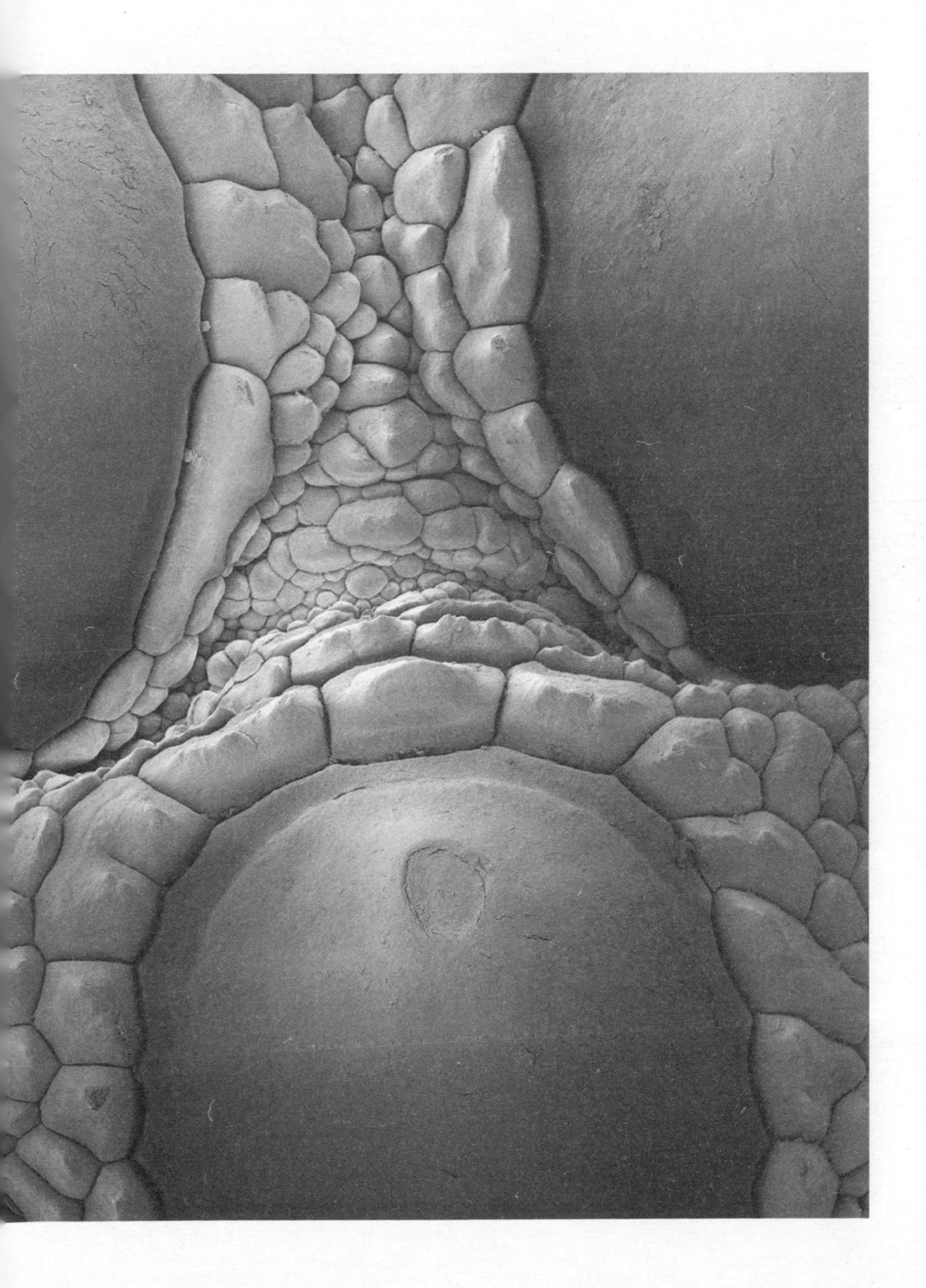

skin impression and find out that, yes, there are melanosomes there and we can actually figure out the colour on these guys.'

The fact that birds are the direct descendants of dinosaurs means that dinosaurs are not extinct. There are some 10,000 species still around. In fact, dinosaurs are actually divided into two groups, avian (those that fly, which we call birds) and non-avian (the land-dwellers that we normally think of as dinosaurs). For the sake of simplicity, when everyone from the layperson to the veteran palaeontologist uses the generic term 'dinosaurs' in conversation, they are generally referring to non-avian dinosaurs.

Dinosaurs were divided into two orders by the British palaeontologist Harry Seeley in 1888: the saurischians and the ornithischians. Seeley characterized these orders (or lineages) by the arrangement of the bones in the hip. The saurischians, which have a pubic bone that slopes down and forward, were named 'lizard-hipped' because their hip structure resembled that of a lizard. The ornithischians, which have a pubic bone that slopes down and backwards, were named 'bird-hipped'. Despite the fact that ornithischians were named bird-hipped because their hip structure was similar to birds, Seeley did not identify any similarity to birds. In fact, further study determined that modern birds actually evolved from the lizard-hipped saurischian dinosaurs, not the bird-hipped ornithischians.

The saurischians include two major dinosaur groups, the sauropods (large herbivores such as *Apatosaurus* and *Diplodocus*) and the theropods (meat-eaters such as *Velociraptor* and *Tyrannosaurus*). The ornithischians include armoured dinosaurs (such as *Ankylosaurus*), horned dinosaurs (ceratopsia), and duckbilled dinosaurs (hadrosaurs). Though scientists have concluded that the oldest dinosaurs were 225 million years old, they do not know how much earlier in time the common ancestor of the lizard-hipped and bird-hipped dinosaurs lived.

As with most areas of the study of dinosaur science, there is also a major controversy about the extinction of what we commonly call dinosaurs. Did dinosaurs die out catastrophically as a result the Cretaceous–Tertiary extinction event that occurred 65 million years ago, or did they die out gradually over a long period of time due to climate changes or environmental forces? There is quite a bit of evidence to suggest that an asteroid hit the earth 65 million years ago and that not only wiped out dinosaurs but a great numbers of other animals and plants as well. Except for a few explainable aberrations, there are no non-avian dinosaur fossils above the Cretaceous–Tertiary boundary in rocks younger than 65 million years old.

Like many palaeontologists, Currie believes that when the asteroid hit natural selection was already at work. By the end of the Late Cretaceous period, temperature shifts on the planet were becoming extreme. He points to the rocks along Alberta's Red Deer River that stretch into Dinosaur Provincial Park as evidence that factors such as climate change were already at work phasing out certain species.

'If you look at 10 to 15 million years before the asteroid hit, you have more than 40 species of dinosaurs in this region,' he explains. 'By 5 million

'There is no black or white in palaeontology, only differing shades of grey.'

years before dinosaurs became extinct, you have about 25 species of dinosaurs. The rocks that were laid down a million or so years before the end of the Cretaceous along the Red Deer River have fewer than a dozen species of dinosaurs. That is telling me loud and clear that there was something else going on to reduce their diversity, and I suspect it was climatic.'

Further, Currie argues, if an asteroid hit and simultaneously wiped out all the dinosaurs, then there should be an abundance of fossils present on the Cretaceous–Tertiary boundary. However, the preserved evidence does not show this. In North America, the fossil record shows that only a few dinosaurs, including *Ankylosaurus*, *Triceratops* and *Tyrannosaurus*, lived until the end of the Cretaceous period. Studying the succession of faunas over a 10- or 15-million-year period shows that species' diversity was dropping off.

'Prior to the asteroid hitting, something else was going on,' Currie reasons. 'It's like everything: as humans what we do is try to come up with a simple answer, but in nature there is not necessarily a simple answer.' He acknowledges that this is just one of many theories on extinction. 'It's very easy to come up with a new theory for dinosaur extinction; it's not so easy to go out and get the evidence. The problem is that it takes years and years of collecting evidence.'

The same can be said about almost all dinosaur theories. There is no black and white in palaeontology, only differing shades of grey. Short of having a time machine to travel back to the Jurassic or Cretaceous periods, palaeontologists must take to the field and dig for evidence, and then hold new fossil finds up against existing ones. Eureka moments are very rare. Discoveries take years to be prepared, studied and then scientifically described before they are presented to the public for further debate. Even under the best of circumstances, they are met with doubting eyes and contradictory theories rather than front-page headlines.

chapter 2

the dino hunters

The special group of palaeontologists loosely known as dinosaur hunters are detectives who spend their lives trying to put together pieces of an ancient mystery that may never be fully solved. The dino hunter is part historian, part explorer, part scientist. He attempts to reconstruct history which dates back more than 65 million years – or 60 million years before the first human – through a combination of research, fieldwork and lab work. To do so he must raise public and private money to fund expeditions, navigate foreign government bureaucracies to gain access to critical sites, endure the harsh conditions of some of the most brutal climates on earth, outsmart poachers looking to make a buck on the black market, and recruit young palaeontologists to assist in the field and the lab.

Dinosaur finds can be traced to the ancient times of the Western Jin Dynasty of China, when they were thought of as 'dragon bones', although the first identified dinosaur find occurred in 1676, in

England. Part of a large bone was found in a limestone quarry at Cornwall and described the following year by Oxford professor Robert Plot as the femur of a large animal like an elephant. However, some people concluded later that it must have belonged to a giant human like those written about in the Bible. It wasn't until 1824 that the first scientific description of a dinosaur was written by another Oxford professor, William Buckland. After collecting a large number of dinosaur bones over a nine-year period, Buckland determined they were all from a related animal that resembled a giant lizard, and he published a scientific paper describing a great fossilized lizard that he named *Megalosaurus*. Although this was the first dinosaur described, the term 'dinosaur' wasn't coined until 1842 when Richard Owen recognized that the remains shared a number of features and therefore should be grouped together taxonomically.

An artist's impression of the duckbilled dinosaur Corythosaurus.

The first recorded North American dinosaur find was made in 1838 by John Estaugh Hopkins in a mudstone quarry on the Cooper River in Haddonfield, New Jersey. Hopkins displayed in his house the bones he found, where they were seen by his friend William P. Foulke, a part-time geologist. Foulke returned to the quarry and discovered most of a skeleton, which he asked palaeontologist Joseph Leidy to help him extract and study. Leidy scientifically described the find and named it *Hadrosaurus foulkii*, or 'bulky reptile'. The *Hadrosaurus* find would lead to what became known as the 'bone wars' which erupted between dinosaur hunters Othniel Charles Marsh and Edward Drinker Cope, who were both mentored by Leidy.

Marsh, of Yale University's Peabody Museum of Natural History, and Cope, of the Academy of

Natural Sciences in Philadelphia, were initially cordial and shared information on their finds. But a dispute broke out over a fossil site that was discovered by Cope just south of the Haddonfield site. Cope was having the site's new finds sent to him for examination, but when Marsh realized there were an untold number of bones to be recovered, he began paying those digging the site to divert the finds to him. Cope found out about Marsh's underhanded tactics, and the gloves came off.

For the next 20-plus years, Marsh and Cope used all means available – including bribery and theft – to outdo one another in the search for dinosaur bones. Both men spent considerable sums of money trying to win the 'bone wars' and to be recognized as the undisputed leader in the new field of dinosaur palaeontology. Marsh was backed by a wealthy uncle, George Peabody, the financial backer of Yale's Peabody Museum, while Cope used his considerable family money. Some of their work was rushed and sloppy. Marsh named both *Apatosaurus* in 1887, and *Brontosaurus* in 1889. He did not have a skull for the latter and when it was mounted, a *Camarasaurus* skull was put on it. Eventually it was discovered that (other than the *Camarasaurus* skull) *Apatosaurus* was the correct name for both specimens. Cope made a similar hurried mistake on *Elasmosaurus*. Despite the duo's lack of professional ethics, they found an exceptional number of specimens and localities for future study. Marsh was credited with discovering 80 new dinosaur species and Cope named 56 – and a competitive stage was set for future dino hunters.

Two of the most famous dino hunters who followed them (but who did not resort to underhanded tactics) were Barnum Brown and Roy Chapman Andrews. Brown, who was named after the legendary circus showman P.T. Barnum, was the American Museum of Natural History's main dinosaur collector in the early 1900s. He is best known as the man who discovered *Tyrannosaurus rex* in 1902 in the Hell Creek Formation in south-eastern Montana. Several years after that landmark find, he travelled to Alberta, Canada and led an expedition down the Red Deer River that discovered, among other dinosaur fossils, several *Albertosaurus* hind feet in the same location, which he had shipped back to the Museum of Natural History.

Andrews was an adventurer who also worked for the American Museum of Natural History and later became its director. He began his natural history quest studying whales in the coastal waters near Vancouver Island in British Columbia, where he collected the skeleton of a humpback whale. He next convinced the museum's director, Henry Fairfield Osborn, to fund the first expedition to the Gobi Desert in Mongolia to search for the origins of man. Though Andrews failed to find any evidence of early man over the course of five expeditions from 1922 to 1930, he and his team found reams of dinosaur skeletons, among them the first

Protoceratops, *Oviraptor* and *Velociraptor*. On their expedition in 1923, they identified the first fossilized dinosaur eggs, which were initially thought to be those of *Protoceratops* but were identified in 1995 as *Oviraptor* eggs.

The work of Brown and Andrews set the standard for future dino hunters. Their finds were remarkable in their time, but they would become even more important as they were re-examined by future palaeontologists. Their work inspired and informed a group of palaeontologists who would revolutionize the discipline in the late 1970s and would ultimately galvanize public interest in dinosaurs.

'The reason that the world needs palaeontologists is because, just think about it: if you go back 68 or 70 million years ago, the largest animal on land was a dinosaur, and mammals were very small,' palaeontologist Louis Jacobs explains. 'If you go to Africa now, the largest land animal in the world is an elephant, a mammal, and you can go and you can see a bird sitting on top of an elephant. That's a dinosaur on top of a mammal. So what's happened in the last 65 million years is the entire ecological arrangement of life has changed, and if we can understand how that ecological arrangement of life has changed and why it changed, that tells us about how the world works. Knowing how the world works is what humans need to know right now to understand the future.'

* * *

Phil Currie wanted to be a dinosaur hunter for as long as he can remember. When he was 11, Currie read Andrews' book *All About Dinosaurs*, and he was hooked. He was captivated by the life led by dinosaur hunters, and he loved reading about how Andrews battled wild animals and travelled to exotic places to search for evidence of these huge beasts, and then returned to the lab to study them. He knew right away what he wanted to do with his life. 'I had a fascination with the lifestyle and the animal, this two-legged beast that looks a little bit like us – but is meaner,' he says.

Currie's first quest came when he was a young, six-year-old boy searching for a plastic *T. rex* in his Rice Krispies cereal box. Each box of cereal had a sticker declaring 'Free Dinosaurs!' Currie managed to collect all the other 'dinosaurs', although some of the plastic figures were not actually dinosaurs – they were flying reptiles or, in the case of *Brontosaurus*, misnamed. In the beginning, Currie would dig into the box for the *T. rex*. When it wasn't there, he would ask his mother to buy another box. But his mother made him eat the entire box of cereal before she bought another. 'By the time the end of the promotion came, I had multiple copies of everything and I was sick of the cereal, but I didn't have the *T. rex*,' he says.

Soon, Currie was looking at the genuine article. He regularly visited the Royal Ontario Museum to see dinosaur remains. At the age of 12, his mother arranged for him to meet Bill Swinton, the museum's director. Swinton was an English scientist who had done some work on dinosaurs and written a number of books on them. He took a liking to the ambitious Currie and introduced him to other palaeontologists. Currie would write to them and ask what classes he should take in high school to put him on the road to becoming a dinosaur hunter.

After several museum visits, Currie realized that most of the dinosaurs in the museum had come from Alberta. In 1961, he convinced his parents to take the family on a trip out west so he could visit actual dinosaur sites. 'It was great because I got to see the field sights where the specimens were being taken out, but it was a little bit disappointing because I didn't see this wealth of dinosaur skeletons I imagined were there,' Currie recalls. 'I figured that if there are so many dinosaur skeletons in the Royal Ontario Museum and so many specimens were found in Alberta, there must be far more to be seen. Well, there wasn't because they were all in the eastern museums.'

Often children become interested in dinosaurs when they first see the impressive re-creations, but over time they lose their interest. In Currie's case, whenever his interest waned he would return to the Royal Ontario Museum and see the real dinosaur skeletons again. By the time he was 12, he had decided that he would move to Alberta and dig dinosaurs when he grew up.

After high school, Currie put himself on a path to becoming a dinosaur palaeontologist. He attended the University of Toronto so he could volunteer in the lab at the Royal Ontario Museum, located on the university's campus. As a college student, he began working on the dinosaur bones the museum was studying, and he made valuable personal contacts in the field.

By the time he finished his bachelor's degree, Currie realized that he needed to leave Toronto because there was no one there working full time on dinosaurs or fossil reptiles to mentor him. He studied for his master's degree at McGill University in Montreal, where noted palaeontologist Bob Carroll became his supervisor. Because Carroll

'Whenever his interest waned Currie would return to the Royal Ontario Museum.'

worked primarily on the origin of reptiles, Currie did not focus directly on dinosaurs, but he knew that he could accumulate a base of knowledge and apply it to his dinosaur studies later.

In 1976, after Currie had started working on his PhD at McGill, he received a call from a former professor of his at the University of Toronto. A job had opened up at the Provincial Museum of Alberta (which became the Royal Alberta Museum in 2005), and the professor encouraged Currie to apply despite the fact that Currie had completed only a year of his doctorate. Although the museum said that it was looking for someone with a PhD, Currie decided to fly to Alberta and attend an interview anyway.

Though other applicants were more qualified Currie was the only candidate who wanted to work specifically on dinosaurs; the others wanted to concentrate on fossil mammals and were only conceding to work on dinosaurs to further their other interests. Luckily for Currie, one of the scientists on the committee that interviewed him was David Spalding, who had a strong interest in the history of palaeontology and subsequently wrote books on hunting for dinosaurs.

'I had no right to get that job,' Currie says. 'There were people that I was competing with who were much better qualified that I was, absolutely no question about it. But when I walked in and did the interview, it was like I became a different person. Subconsciously I was telling myself that this was my job and that I was the best person for it.'

Currie was hired, and with that he became one of a select group of full-time dinosaur palaeontologists in North America. The museum wanted Currie to undertake fieldwork, stage exhibitions and handle education and research. Currie couldn't have been more motivated. But when he arrived, he discovered that his annual budget was a scant $4,000 – less than the cost of collecting a single dinosaur find. The struggle for research and fieldwork funding began immediately.

Dinosaurs are big and therefore logistically expensive to remove, and for years the financial resources weren't readily available. The Second World War had wiped out money for most scientific research, and when funding recovered it was funnelled first into medical sciences. Though the public was hungry for information about dinosaurs, the funding simply wasn't there. Research money went first into fossil fish, small fossil mammals, and anything connected to human origins and the 'missing link'. For years, palaeontology found itself at the end of a long line, and by the late 1970s there were no more than a dozen full-time dinosaur palaeontologists in the entire world.

'The large museums like the American Museum of Natural History and the Royal Ontario Museum felt that they already had their wonderful dinosaur displays,' Currie says. 'They felt they didn't need any more, so why should they hire a dinosaur palaeontologist to do it? But the public still loved it. That kind of worked against us too – other scientists would say dinosaurs are popular in the public as kind of a gee-whiz thing, and why should we do research on dinosaurs when they are extinct?'

But the dinosaur renaissance of the late 1960s and early 1970s changed things. Leading palaeontologists John Ostrom and Dale Russell, followed by Ostrom's students, Peter Dodson, Jim Farlow and Bob Bakker, were all working on ideas concerning dinosaurs as living animals and discussing them in terms the layperson could understand. 'Things started to turn around when you started to get a few palaeontologists who were studying dinosaurs and made some rather special finds,' Currie says. 'The finds weren't all that unique, but the way they marketed them was different. These were biologists rather than geologists, so they were looking at things differently.'

As the dinosaur renaissance was getting into its stride, Currie arrived at the Provincial Museum of Alberta in 1976 and staged his own mini-revolution in dinosaur research in Canada. Despite his limited budget, Currie dedicated himself to fieldwork. The odds were against him. The Alberta museum had been looking for years and hadn't found anything interesting. Many palaeontologists believed that the eastern museums had already taken all the good specimens and there were none left.

There was also the issue of time and manpower. Dale Russell had done a study of the field notes of dinosaur hunters like Barnum Brown working in Alberta from 1910 to 1924 and calculated that, on average, during the golden period when major dinosaur discoveries were being made, it had taken about four man-months of walking to find one good specimen. Therefore, if you had four people searching, it would take one month to find a dinosaur. Currie had just one other person helping him and they would be in the field for only a month. What they needed was some luck.

Currie forged ahead with great determination. He knew early success was critical, and he had some. In his first year, near the Canada–US border,

Currie collected a hadrosaurid, or duckbill dinosaur. Based on fossilized footprint evidence, duckbills walked around mostly on their hind legs and would go down on all fours whenever they were feeding. The biomechanics of their bodies also indicated that they weighed too much to support themselves full time on their hind legs. They had very long skulls, and at the front of the skull the mouth expanded into a duckbill. A hadrosaurid didn't have teeth in the front part of its mouth but it had many in the back part of it, a structure that worked well for eating plants. Currie's find generated enough publicity that people began to know who he was. He went way over budget on his first field outing and, had he not found something noteworthy, he says, he probably would have been laid off.

The following year, a petroleum company drilling a pipeline in Canada's prairie badlands found another duckbill dinosaur. Currie convinced a government official to give him some additional summer funding and he started a volunteer programme so he had enough people to help excavate it. Together, Currie and his two technicians, along with four university students and one high-school student, headed to the badlands to dig out the dinosaur.

'It was a good specimen – the part that was exposed,' Currie says. 'But as we followed it into the hill, literally moving tons of rock, and we got to the front, suddenly I found one tyrannosaur tooth in the ribs. I thought, uh oh. Then we went further and found another tooth. Then we kept going and found more tyrannosaur teeth and broken bone. So the front end of the skeleton had been eaten off by a tyrannosaur, probably *Albertosaurus*. The tyrannosaur had bitten through the bone. So we ended up with fragments of the hadrosaur bones and the teeth from the *Albertosaurus* where it had munched. That was kind of cool, and it did attract attention. It started a cycle.'

By 1978, Currie and his team were on a dinosaur-finding roll, so much so that the museum eventually ran out of storage space. Currie also arranged an exhibition called 'Discovering Dinosaurs', for which

'Currie always had it in the back of his mind to create a dinosaur display.'

he borrowed all the Alberta dinosaur finds from other museums. 'Discovering Dinosaurs' broke all attendance records in 1979 and generated reams of publicity for the museum.

The Provincial Museum of Alberta was a government institution and did not accept funding from private sources. The publicity from the finds and the exhibition caught the government's attention and resulted in officials increasing funding for dinosaur hunting. However, for many years, even though a budget had been established for dinosaur research, when Currie reached the end of the fiscal year he would discover that the funding had been used to plug a hole in the budget somewhere else.

'The politicians were concerned about dinosaur finds in Alberta,' Currie says. 'Our task was to prove that we still had the fossil resources and all we needed was manpower and money. We had to work a little bit on people's pride because it is kind of a strange thing to know that you are really rich in something but to see it you have to go to another part of the world. We played a little bit on both things.'

Alberta was famous for its dinosaurs, yet it had no centralized place to showcase them. Currie always had it in the back of his mind to create a dinosaur display worthy of the dinosaur resources. At the Provincial Museum of Alberta, he had less than 45 square metres (500 square feet) of space to showcase his finds. 'It was pathetic that all the biggest dinosaur Alberta displays were somewhere else.'

His big opportunity came in 1979 when Dinosaur Provincial Park in south-eastern Alberta had become such a hotbed of finds – led by Currie discovering and excavating a *Centrosaurus* bone bed there – that it was named a UNESCO World Heritage Site. Dinosaur Park, as palaeontologists know it, covers nearly 19,000 hectares (73 square miles) and has produced more individual skeletons of different dinosaurs than any place of its size in the world. Currie began putting together a planning team to build a museum to showcase Alberta's dinosaurs that was worthy of a UNESCO site.

In 1980, with the full support of Alberta Provincial Premier Peter Lougheed, the government agreed to finance a dinosaur museum. However, the museum would not be located in Dinosaur Park, because of the sensitive nature of the site, but nearly 170 kilometres (105 miles) away in Drumheller, a small town of 7,000 located in the heart of Canada's badlands. Drumheller had been marketing itself as the land of dinosaurs since the 1950s. The town had a small dinosaur museum and several dinosaur-themed parks.

Currie was asked to submit a proposal for the museum. Knowing that government officials were generally more concerned with the public aspect of a museum than the research side, he outlined several options. One was to put in an interpretation centre in Drumheller – essentially a display without the back-up of a scientific research team on site. For the science, the museum would use palaeontologists in either Calgary or Edmonton, big cities with university-based possibilities for research.

'I wasn't keen on that because the reality is that displays like that tend to be dead displays in the sense that they are static. All the information comes in second hand, and there's no real incentive to get things going,' Currie says. 'I put the hard sell on: "If you want to do this, you should do it right. You should store collections, have research staff to take care of the material and to collect new specimens, educational staff and the whole works; then you have a living museum." I also knew the power of publicity. There are two kinds of publicity. One is the kind you pay for and it's very expensive. As a government organization, we obviously didn't have the budget for that. And then there is the kind of publicity where you do research and, if it's interesting, it attracts the media and lets people know you exist.'

On Christmas Eve 1980, Currie received a phone call and was told the project was going to be the 'Cadillac version'. In addition to the galleries, it would have a library, study centre and laboratory facilities. Currie made sure the museum had an extensive library, which at that time was a key to attracting top scientists, and a massive storage area that could house the specimens collected and awaiting examination. He was also able to persuade the government to lift its normal rules concerning scientists' salaries, support staff and facilities. The museum was budgeted at $42 million though, because it was built during the 1981 recession, the final price tag was only $28 million.

The museum was completed in 1985 and was named after Joseph B. Tyrrell, who had arrived in 1884 on assignment to conduct a geological survey of the Red Deer River. He travelled to the badlands surrounding Drumheller looking for coal, and he accidentally stumbled on the skull of a large carnivore that was later named *Albertosaurus*. The museum would focus on celebrating the amazing finds that lived in Canada's prairies more than 65 million years ago.

On its opening weekend, the Tyrrell Museum drew 30,000 people, and by the end of the first year, it had recorded 500,000 visitors. The dinosaur museum has remained so popular over the years that it has created its own economy and now attracts over $15 million in tourist revenue to the area annually. In 1990, the 'Royal' epithet was added, making it one of the most distinguished museums in Canada. 'Suddenly, tourism started to be talked about in Alberta, and now it's one of our major industries,' Currie says. 'We have people come to Alberta from Europe or Japan and parts of Asia who come specifically to see the Tyrrell Museum.'

Once the Tyrrell Museum was up and running, Currie was itching to return to the field. He had decided that he did not want to be a museum director. He felt that supervising the building and staffing of the museum would slow his progress as a dino hunter. An interim director supervised construction while a full-time director was sought. Currie served as an assistant director so he could maintain some control over what happened in the research and the collection sides of the museum, but once the museum was fully up and running, he backed out of that role and became nominally attached as Curator of Dinosaurs. 'I wanted to go out in the field and do research and not be tied to the building giving tours, doing the administration and taking care of people,' he says.

In October 2005, he left the Royal Tyrrell Museum to take up the Canada Research Chair in the Biological Sciences Department at the University of Alberta. For him, it was the ideal situation. He would have a base at the university which provided him office space and a laboratory, and he would be free to spend four months a year in the field searching for dinosaurs.

For Currie, fieldwork has always been the most exciting part of his craft, particularly when he and his team came up with major finds. 'In Alberta, we excavated two *Tyrannosaurus rex* and I was very lucky that I was able to do that,' he says. He laughs at the irony: as a child, he never did get the *T. rex* action figure in his Rice Krispies, but he found not one but two in real life. The most recent *T. rex* find was in Saskatchewan, the next province to Alberta.

But Currie has since turned his attention to *Tarbosaurus*. *Tarbosaurus* is an animal that initially was described as *Tyrannosaurus bataar*, and the relationship between the two predators fascinates Currie. One of the tyrannosaurs from Alberta is an animal called *Daspletosaurus*, and Currie says that it appears that the *Daspletosaurus* is probably the direct ancestor of *Tyrannosaurus rex*, and possibly also of *Tarbosaurus*.

'One of my very first discoveries of a tyrannosaur was a specimen of *Daspletosaurus*,' Currie remembers. 'I had been walking around the badlands in Dinosaur Provincial Park in southern Alberta when I stopped to take a picture of the panorama of badlands, and my camera case fell off and rolled down the hill. I went down the hill to pick up my camera case and was very shocked to see that it had landed on a skull of *Daspletosaurus* – though I didn't know that at the time. But a couple of years later we had excavated the whole skeleton and I realized that it was, in fact, *Daspletosaurus*.'

Looking for bones requires extreme focus. Currie says that generally when he arrives at a new site, he finds that it can actually be difficult to locate dinosaur bones at first. He believes the reason is that he has a search image that develops over time. Basically, this image allows him to think about other things when he's searching for dinosaur bones.

'You can walk along and think about your last supper or your dream the night before, but as soon as you see a dinosaur bone – bang – your whole attention gets focused on that,' he explains. 'And so the search images are a very important aspect of hunting for dinosaur bones. Now when you go to a different region, the trouble is that the bones are in a different kind of rock; they can have a different colour, and different textures. You generally find that until you discover your first specimen, you haven't got that search image developed, and so you can walk right over a good dinosaur specimen because you're not really looking for that; you're looking for something from another part of the world at a different locality. Generally, of course, if you've been hunting these specimens for a long time, it takes less time for you to switch from area to area, but there is this problem with search image for sure.'

Even for experienced palaeontologists, finding dinosaurs requires a combination of science and serendipity. 'Some people are lucky and some people aren't,' Currie says. 'Generally what it comes down to is you have to be the right person in the right place at the right time. If my camera case hadn't rolled down and I didn't know what one of those specimens looked like, then I wouldn't have seen it. So if you put yourself in the right situation, right time, right place, and you know what you're looking for, you are going to find these things. But there is always a certain amount of luck.'

Dino hunters draw on the work of their fellow palaeontologists. It's a small community, and although it is very competitive – sometimes bordering on the cut-throat – Currie refers to fellow palaeontologists as his 'colleagues' even though they generally work independently. For him, working together is essential because the subject matter is so dauntingly vast.

'Sometimes it is a matter of finding animals that washed in from a different environment 65 million years ago; or maybe a new species represents an animal that was just not very common in that particular area in the Cretaceous, and therefore it didn't fossilize very commonly; in other cases it is simply because the environments that they lived in just aren't represented.'

In 2006, Yuong-Nam Lee, a Korean herbivorous dinosaur specialist and expert dinosaur hunter, was able to put together financing for the Korea–Mongolia International Dinosaur Project, a five-year expedition in the Gobi Desert. He has been studying dinosaurs for 15 years, and serves as Principal Researcher (and now Director) at the Korea Institute of Geoscience & Mineral Resources in South Korea. He had hoped to work with Currie on an extensive expedition for a long time. 'Dr Philip Currie is a famous dinosaur person all over the world,' Yuong-Nam explains. 'His knowledge is incredibly huge, so I learn many things from him.'

He and Currie would be joined by a group of scientists that they had known and worked with in the past. Each had their own area of expertise and working together would allow them to pool their resources. Yuong-Nam's close friend Yoshi Kobayashi is Japan's number one dino man and an expert on theropods – meat-eating dinosaurs. Yoshi has been working in the Gobi since 1996.

Louis Jacobs, Professor of Earth Sciences at Southern Methodist University in Dallas, Texas, served as PhD advisor to both Yuong-Nam and Yoshi. Jacobs studies vertebrate fossils to determine what they tell us about the history of the Earth and the history of life. He tries to go to places where that record can be improved and where there's a chance of making breakthrough discoveries and putting them into context so we can learn more about past life on the Earth.

'The main reason I like doing [fieldwork] is because that's where you find fossils, and palaeontologists have to have fossils,' Jacobs says. 'It's very much like a different world being there, but it really was a different world at the time those dinosaurs lived, completely separated by a distance of time. And the only way that you bridge that gap in time, that huge barrier, is to understand what happens each step of the way for the last 70 million years. It's uncanny when you think about it, that you even go to a place like [the Gobi Desert] and find the kinds of fossils you do, and ask the sorts of questions that you can ask, and understand it in terms of what the world is like today. But it's those general principles that will allow us to really understand the Earth, and that's important for humans to know.'

Dave Eberth, a Canadian sedimentologist and habitat specialist also joined the expedition. Eberth is a senior research scientist at the Royal Tyrrell Museum of Palaeontology, and is basically a geologist who works with fossils. He is able to look at how rock formations came together millions of years ago, and then use his knowledge to create a bigger picture of what life was like at the time.

Currie met Eberth at a Society of Vertebrate Paleontology conference. He was so impressed that he hired Eberth to work at the Royal Tyrrell

Museum. The two have a feisty rapport. Indeed, Eberth has often said he is sceptical of Currie's theories about dinosaur behaviour. 'Anything I say, he objects to; anything he says, I object to,' Currie says. In their case it's not totally contentious; it pushes each other to solidify their theories and move the scientific dialogue forward.

These world-renowned experts debated and studied the merits of Currie's theories about carnivores. 'Having a team like this, where everybody has come from different experiences, different parts of the world and has a different perspective on things, means that you can sit down, talk and brainstorm and, in fact, come up with ideas that you probably wouldn't have developed if you'd just been by yourself,' Currie says.

He likens the group to a construction team. 'We're in there building a house and putting it together, and we're starting from the foundation,' he explains. 'The foundation, of course, is what gets excavated in the field. As we're developing things, somebody's got the carpentry expertise, somebody's got the wiring expertise, somebody's got the plumbing expertise. By the time you get to the end of the process, you've got something you can live with. When you're working on a big multinational project like this, it's just a wonderful aspect of it because you get to balance ideas.'

After the expedition, Currie would take the information he gathered to the foremost lab technicians. He would call on Larry Witmer at the

Ohio University College of Osteopathic Medicine. Witmer, who is a professor of anatomy and runs the Witmer Research lab, uses the latest CT scanning equipment to research dinosaur brains to determine the quality of their sensory perception – critical in helping complete the emerging picture of dinosaur behaviour.

Currie's evidence would also be examined by John Hutchinson at the Royal Veterinary College in Hatfield, England, who has developed new ways of estimating animal running speeds. Hutchinson's techniques, which use a special motion-capture process involving computers, are fundamental to figuring out how fast tyrannosaurs could run.

In his quest to study tyrannosaur behaviour, Currie would test and re-test his theories. He would return to the field to dig for additional evidence, spend time in the lab examining the evidence and employ cutting-edge techniques of science to study the ancient beasts.

'Palaeontologists are kind of like detectives in a way; you go back to the scene of a crime and you look at all the evidence and you try and interpret what's going on,' Currie says. 'It's a wonderful mental exercise.'

chapter 3

the badlands of canada

anada's badlands are one of the world's greatest sources of dinosaur bones. The badlands feature a seemingly endless open sky that hovers over grassy prairies leading into canyons with multicoloured layers of rock and a series of rivers slicing through them. The sediment layers where so many dinosaurs have been found were exposed at the end of the last ice age some 15,000 years ago by rapidly melting glaciers that stripped the landscape and created the exotic-looking rock formations that are at turns beautiful and eerie.

In the Late Cretaceous period, some 65 to 100 million years ago, Alberta had a climate that was similar to present-day North Carolina. It was a humid coastal plain containing a dense subtropical forest of palms and giant redwoods. Water flowed from the Rocky Mountains to the west through marshes, swamps and forests, and emptied into an inland sea that once submerged much of the central part of North America. At that time, Alberta was home to fish, turtles, salamanders, crocodiles, small mammals, and lots and lots of dinosaurs.

Over the years, more than seventy different species of dinosaurs have been found in the badlands, particularly in the areas known as Dry Island Buffalo Jump Provincial Park, located about two-and-a-half hours north-east of Calgary by car, and Dinosaur Provincial Park, which is about two-and-a-half hours' drive south-east of Calgary. Half of the dinosaur skeletons discovered here are hadrosaurs, the duckbilled dinosaurs. They lived alongside horned ceratopsians, armoured ankylosaurs, domed-headed pachycephalosaurs, bird-mimic ornithomimids, swift dromaeosaurs and big carnivores like *Gorgosaurus*.

Many of the amazing finds are housed at the Royal Tyrrell Museum of Palaeontology in Drumheller, which has dubbed itself 'the dinosaur capital of the world'. The museum's highlights include two *Tyrannosaurus rex* specimens on display, one of which comes from near Dry Island. There is a very beautiful *Gorgosaurus* from Dinosaur Park, a small cousin and ancestor of *Albertosaurus*, the dinosaur that is such an integral part of Phil Currie's work.

One of the dinosaurs on display at the museum is *Pachyrhinosaurus*, which was first discovered in the Alberta badlands in 1946. Fossil hunter Charles M. Sternberg described and named *Pachyrhinosaurus* in 1950. A member of the ceratopsian family of horned herbivorous dinosaurs, *Pachyrhinosaurus*,

RIGHT
The Albertosaurus *bone bed in Dry Island Provincial Park, looking towards the ridge that helped Currie rediscover the site in 1997.*

BELOW
A copy of the photo Barnum Brown took in 1910 of Peter Kaisen working in the Albertosaurus *bone bed.*

or 'thick-nosed lizard', was named for the rough, thickened bones of its nose. Not much was known about the life-cycle of this dinosaur until a massive bone bed was found in 1972 at Pipestone Creek in northern Alberta that contained hundreds, possibly thousands, of *Pachyrhinosaurus* at various growth stages. To the scientists who examined the bone bed, the mass-death event suggested seasonal herding behaviour.

Currie cut his teeth as a dinosaur hunter in Canada's badlands, and it was his finds there that started him thinking that dinosaurs may have lived in gangs. Once Currie became ensconced at the Provincial Museum of Alberta in 1976, he reviewed many of the previous Alberta finds. His aim was to determine which sites had not been fully explored or excavated.

One of the first finds he reviewed was a footprint site that had been discovered in 1924 in Peace River Canyon in British Columbia and written about in a scientific paper in 1930. For years nobody paid much attention to the site because so few people were actually working on dinosaurs. However, there was now a need to return to the site urgently. A hydroelectric dam was being built on the site, and upon completion it would flood the entire area, inundating the footprints that had been made more than 100 million years earlier.

Currie had to act quickly. He contacted the hydroelectric company and convinced it to fund his expeditions as a way to give back to the community.

This would allow him to immediately begin studying the footprints before the land was completely flooded. Currie and his team worked there for four field seasons. (Field seasons are in the spring and summer and last up to a month, as the canyon is buried under heavy snow in winter.) However, as much as big business was helping him, it was also pushing him out of the way.

'To put it in perspective, the hydroelectric company actually gave me a budget to work for that one month of the year that was about five times the size of my budget for the rest of the year here in Alberta,' Currie recalls.

However, he and his team were working downstream from a different dam that had been built in the 1960s by the same company. Every time Vancouver needed power at suppertime, the company would release water from the dam to generate the additional electricity. Currie appealed to the company to curtail the release of the water because the run-off was flooding the sites where he was working. He pointed out that the company was giving him $30,000 to do this work – and then flooding the site. The response? 'It doesn't matter because the $30,000 we are giving you we make in less than an hour when we turn those generators on.'

The footprint studies that were completed before the dam was finished and the entire site flooded in 1980 proved tremendously valuable examples of behavioural trackways. The Peace River Canyon site contained literally hundreds of trackways that provided clues in terms of how the dinosaurs moved. 'We started thinking about theropods as being gregarious animals when we started looking at dinosaur footprints,' Currie says.

What was most intriguing about the trackways was the fact that very often Currie and his group found not just one trackway but multiple trackways, all going in the same direction. The parallel trackways at one site followed an S-curved course, and by studying them Currie concluded that the dinosaurs were in fact moving together side-by-side, and that there was some kind of gregarious, or gang, behaviour at work.

'There was a set of trackways from a plant-eating dinosaur in particular that made a real impact on me,' Currie recalls. 'There were four animals that

'Currie's group found not just one trackway but multiple trackways, all going in the same direction.'

were obviously walking side-by-side and, at one stage, one of the animals slipped. It was so close to the next one that, when it shifted sideways, it bumped into it. You can see its trackway suddenly taking this nice little deviation, not quite as much as that of the first one because he didn't slip and fall, but he moved sideways and forced the next one over to move. The third one moved sideways and forced the fourth animal to deviate as well. There is absolutely no question that this long series of trackways – as many as 50 footprints long – show that these animals were moving exactly parallel to each other and were side-by-side. There was no way of escaping the fact that these were gregarious animals.'

It was obvious to Currie that the plant-eating dinosaurs were moving in groups. He had traced up to 17 animals moving side-by-side across the big mud flats. But even more interesting was that mixed in with the herbivore footprints were carnivore tracks. The carnivores were animals of about the same size, a little bit bigger than *Ornithomimus*, which was 3.5 metres (12 feet) in length and weighed between 100 and 150 kilos (220–330 pounds), but smaller than the 9-metre (30-foot) long, 2 tonne *Albertosaurus*. Though scientists haven't been able to identify precisely the animal, they know it was much older in age and existed before tyrannosaurs inhabited the area. However, with the carnivores Currie noticed a different pattern to the footprints:

'The big carnivorous dinosaurs were moving in groups as well, but they weren't moving side-by-side. They were kind of criss-crossing and moving among each other, possibly tracking the plant-eating dinosaurs in some cases, or that is the way it appeared from the trackway sites. So we had very strong evidence that the plant-eaters were gregarious, and we had some evidence to suggest that the carnivores were also moving in small groups or packs to hunt those plant-eating animals.'

Based on this trackway evidence, Currie began to theorize that if herbivores were gregarious then it was likely that carnivores were, too. 'After all, if the herbivores are moving in big groups then it makes sense that the carnivores would also respond in kind; the carnivores would basically have to pack to hunt herds of herbivores,' Currie explains. 'It is the same kind of pattern we see today in many parts of the world: where herbivores move in big groups, carnivores do it as well.'

This type of behaviour is at work on the modern-day African plains, where carnivores such as lions move in prides to hunt, while herbivores such as zebras move in herds to protect themselves. 'Large herds tend to be in open environments,' Currie says. 'As you get herding herbivores, they can warn each other about carnivores coming and form groups to face the carnivores and keep them at bay. At that stage, what happens is that the carnivores start grouping and start practising social behaviours so that they can hunt.'

By the time Currie had collected the trackway evidence, he had also uncovered bone beds of the herbivorous ceratopsians in the Alberta badlands. 'They were indicating the same thing: gregarious behaviour.'

But he knew that he needed to find and study bone beds for carnivores to advance his theory. For years, it was assumed in the palaeontological world that tyrannosaurids – the two-legged, small-armed carnivores like the *T. rex* and its cousin, *Tarbosaurus* – were solitary hunters. The evidence to prove this theory was very simple: only single skeletons had ever been found, hence they must have lived and hunted alone. There had been evidence to show that smaller meat-eaters like *Velociraptor* – the Mongolian dinosaur made famous in the *Jurassic Park* movie – hunted in groups. But few people believed that the big 'kings of the land' behaved this way.

Digging Dry Island Bone Bed, rediscovered by matching the ridges and trees in the background.

Currie's next important find came in Dry Island Park, the most important *Albertosaurus* bone bed in the world. It was first discovered by Barnum Brown

around the turn of the twentieth century and rediscovered by Currie in 1997. He found it through the classic palaeontological combination of detective work and physical labour.

In the early 1900s, Brown worked with the American Museum of Natural History's director, Henry Fairfield Osborn, and collected many of the famous dinosaurs on display in New York. He worked originally in Wyoming and other parts of the United States, and first went to Alberta in 1909. The impetus for Brown to travel to Canada came from an Alberta rancher who had visited the museum in New York and seen all the dinosaurs on display. The rancher went to the curator and told him that on his ranch, located just outside where Drumheller is now, he found dinosaur bones all the time. So in 1909 Brown went to Alberta to see what the possibilities were for collecting dinosaurs. He saw so much bone that he decided he would mount a proper expedition.

Brown returned to Alberta in 1910. He built a boat in the city of Red Deer and floated down the Red Deer River. Initially, he was accompanied only by his assistant, Peter Kaisen. As they floated downstream they stopped every so often to look for fossils, which they occasionally found. They eventually reached a point in the Red Deer River where they found what Brown believed was an *Albertosaurus* skeleton.

Brown and Kaisen began excavating the skeleton, which was embedded in hard rock. By the second or third day, they realized it was not one animal but several. In the end, they collected what were parts of nine skeletons. They took all the material and put it on the boat and continued down river. By then, they had been joined by a third person who came with a cart to haul their finds out of the badlands and ship everything back to New York City.

'One must assume that Brown knew it was an important site and there was something about it that was really strange, and he intended to publish on it,' Currie says. 'The problem was that as Brown drifted further and further down the river, he got into the richer areas for fossils. Consequently, he ended up collecting so many dinosaurs that he was occupied full time for the rest of his career describing these dinosaurs, as well as doing other things. So he never got around to describing that first find, those *Albertosaurus* bones.'

However, Brown did write one sentence in an article he published in the 1930s about finding a bone bed of carnivores on the Red Deer River. Unfortunately, he never disclosed the exact location. Currie was aware of the existence of the bone bed, and he had searched for it in a cursory manner on his trips down the Red Deer River, but he had so little information to go on that he didn't have much chance of finding it. 'If you were lucky and you happened to run across a bone bed where it's all carnivores, then you would wonder if it was Brown's bone bed, but that's as far as you could go. No one did.'

In 1996, Currie was in New York City for the annual Society of Vertebrate Paleontology meeting. He arranged for himself and Eva Koppelhus to visit the American Museum of Natural History to look at the dinosaur bones in storage that had been collected in Alberta. In a drawer in the basement of the museum, Currie found the left foot of an *Albertosaurus*. He could immediately tell from the preservation that it was from the Drumheller area because it had a certain colour and shine to it that you only find there because of the minerals in the ground. 'Cool, nice ... a gift,' he thought.

Currie opened the next drawer and there was another *Albertosaurus* foot. Then he opened the next drawer and there were two more feet. 'By now, things are started to fire off in my mind that this is probably the stuff that Brown collected from that bone bed,' Currie recalls. 'They were all left feet, different sizes. At that point, I didn't know they were all from the same site. My suspicions were that this probably was the bone-bed material. When you talk about bone beds, most of the time you are talking about animal skeletons that have fallen apart, and all of the bones have been mixed up, so you can't tell which bones belong to which individual. But Brown wasn't collecting disarticulated bones. He was collecting partial skeletons, where you had complete feet with all the bones together and a tail as well.'

Currie foraged though the rest of the drawers in that section and kept finding more bones. Though they weren't all feet, there were a lot more feet than anything else. He and Koppelhus then checked the records and found out that, sure enough, these bones had all been collected by Barnum Brown in 1910 from one site. 'I realized that if he was collecting mainly feet, it's either a really weird bone bed where everything else is washed away and he has nothing but feet. Or, more likely, he was just collecting feet because he wanted to figure out how many animals were in there, and then he left the rest of the bones behind,' Currie says.

As Currie studied the bones, he came to another startling realization: the bones were mostly *Albertosaurus* bones mixed with very few bones of other dinosaurs. 'The *Albertosaurus* material greatly outnumbered all the other dinosaurs,' he says. 'These dinosaurs in a normal population make up about one in 20 animals. So you would expect in any normal situation to find 19 other animals for every *Albertosaurus*. This was exactly the opposite count that we were getting.'

Currie and Koppelhus went into the museum's archives and dug out Brown's field notes. For several hours, they pored over the notes in the hopes of learning the exact location of the find site. They extracted what information they could, though there wasn't very much to go on because Brown didn't keep detailed notes. They then turned to Peter Kaisen's filed notes, but again, there wasn't very much information in them either. There were some clues suggesting that Brown and Kaisen weren't collecting everything they found; rather they were just collecting certain parts of the skeletons. This gave Currie hope that he could re-find the site and get more out of it. He then found four photographs from the trip. The real detective work was about to begin.

The following summer, Currie and a team retraced Brown's journey down the Red Deer River. The joint expedition was funded by a non-profit organization called the Dinamation Foundation, based in Colorado, and the Tyrrell Museum. Using the best two photographs as a guide, Currie searched for the site of the *Albertosaurus* bone bed.

The group reached what they believed was the general area, but they soon realized that the two best photographs had been mislabelled. Now, Currie had to rely on the other two black and white photographs, one showing Brown's campsite by the river and the other of Peter Kaisen sitting on a pile of rubble in a quarry.

The badlands can be very confusing. The rocks layered with orange, brown and red and the undulating hills all look the same, and there are very few distinguishing landmarks. Currie had his work cut out. He studied the photograph of Kaisen in the quarry and noticed a ridge of spruce trees in the background of the 87-year-old photo. He knew that if he could somehow find that ridge of trees – provided they were still in the same formation – he could find the site.

On the first of the four days that Currie and his group were going to spend in what they believed was the general area of the bone bed they had no luck finding the ridge. There was a sameness to everything in the landscape. Like a forensic detective, Currie re-examined the photograph and re-thought his approach. 'I remembered that all these spruce trees are on the north-west facing slope and so that maybe this cliff face is facing east rather than north-west, so I started looking in the opposite direction,' he recalls.

The following day was the hottest of the summer. The mercury hit 43°C (109°F) and by mid-afternoon the group was completely exhausted. They decided to return to their base camp, which was located on an island in the middle of the river. Currie, however, was sure he was headed in the

'The mercury hit 43°C (109°F) and by mid-afternoon, the group was completely exhausted.'

right direction, so he decided to press on for another hour on his own. Walking the deep canyons in comfortable weather is difficult enough, but this trek was brutal and Currie was out of water.

'I must have been about a half a mile away from [the site] when I started thinking, "That is beginning to look familiar." When I finally topped a ridge and saw the trees in a row, I knew I had found the site. At that point, it was about 4 p.m. I had just gone through the hottest part of the day and I was into heat exhaustion at that point. I got excited and I took a new photograph with my camera so I knew I had found it.'

Currie then had no choice but to head for camp and return to the site the following morning. When he reached the river, he tried to bend down to take his boots off for the walk through the water to the island camp, but he couldn't move. His legs had seized up due to cramps because of heat exhaustion. He simply walked into the river with his boots on and headed for the camp. He arrived and found everyone sitting in a circle. Serendipitously, his wife was in the middle of reading aloud the letters that Brown had sent back to Henry Fairfield Osborn talking about the *Albertosaurus* bone bed that Currie had just re-found.

A Tyrannosaurus *maxilla (upper jaw bone) is being prepared in the laboratory on one side. On the bottom can still be seen the plaster jacket that protected it when it was brought in from the field.*

When the group returned to the site the following morning, they found *Albertosaurus* bones all over the ground where Currie had been standing when he took the picture. There were even parts of skulls and teeth. He had to laugh. 'I had been standing right there and I was so caught up with having found the right landscape and so heat exhausted that I didn't look down,' he recalls.

To verify that they had the correct site, one of Currie's technicians removed a piece of rock with the impression of a bone. He then sent it to New York City with a friend who went to the Brown collections in the American Museum of Natural History and found the piece of bone that fitted into the impression in the rock. The museum catalogued the piece of rock that went with the bone that had

'Black Beauty' is the skeleton of a Tyrannosaurus rex *that was collected in the Crowsnest Pass of south-western Alberta in 1982.*

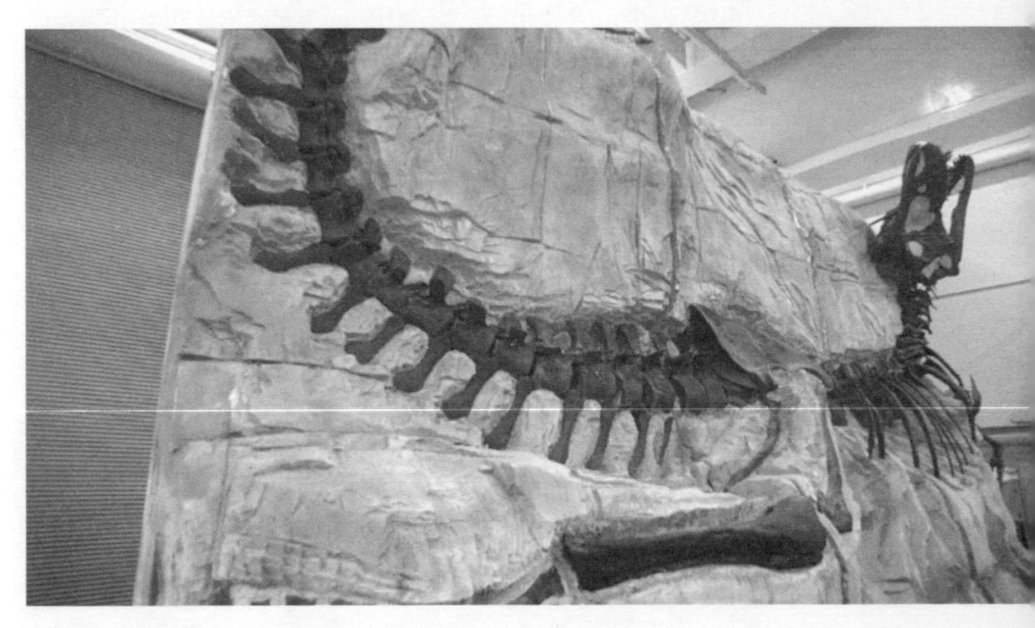

been found almost 90 years earlier, and it confirmed that the two matched.

Because plans for the summer had been made in advance, Currie didn't start excavating the site until the following summer, in 1998. The bone bed was huge, extending half a kilometre in one direction. Brown had clearly left the bones because he did not have the manpower to fully dig the site. Excavations began and confirmed what Currie had thought when he first saw the bones in the museum basement: the bones were mostly *Albertosaurus* bones.

'In a normal bone bed, all the bones are mixed up,' Currie explains. 'This was 85 to 90 per cent Albertosaurs. The latest high estimate of numbers of individuals is 26. We estimate the numbers in a lot of ways. If you are really conservative and you just estimate foot bones the way Brown was doing, then we get to a dozen animals. If the foot bones are all from animals between 4 metres and 9 metres long [13–30 feet], but you find another bone from the skeleton of an *Albertosaurus* that isn't a foot bone but it happens to be from an animal that is larger or smaller than the 4- to 9-metre range, then you add another animal to the count. Using the size distribution of the bones, the minimum number of individuals of *Albertosaurus* could be as high as 26, and even if you are very conservative and use only foot bones, then it is still more than 12.'

The fact that as many as 26 skeletons were found together in such a high concentration suggests that these carnivorous dinosaurs died together. Currie analyses the find. 'It tells us that there is something seriously wrong here in terms of normal ecosystems,' he explains. 'You don't get that many carnivores without a reason. There has to be a reason all of those animals are together. Of course, there are many possible explanations as to why they all ended up in the same place. They all died together. I think we can all agree on that because the preservation of the bone is all the same and the environment, when we look at the geology, is saying these bones were buried over a very short period of time. It's not a long-term accumulation like the La Brea Tar Pits [in Los Angeles] where you could be looking at thousands of years. This is a different game. So we can agree on the fact that animals died together.'

The more important question is did they live together? Were they just migrating? Were they like bears that congregate to eat when food is abundant? Or were they together because they were hunting cooperatively? 'Why they are together is the controversial part of it,' Currie admits. 'Of course, I believe that it is the best indication we have that these animals are in fact gregarious. The reason I say that is because they died together. You are talking about a large number of old and young animals. The ages range from 2 to 24 years old and the sizes varied from about 2 metres [6½ feet] long to almost 11 metres [36 feet] in length. The biggest one is almost as big as a *T. rex*; the smallest one is 2 metres long. That's pretty significant because you can calculate how many predators can be sustained within one area. When you are talking about large predatory animals, they need a huge home range. You would normally expect the hunting territories for that many Albertosaurs, if they are single animals, is going to be absolutely enormous. Each one would need 4 or 5 square miles [1035–1300 hectares].'

One of the mysteries is that Currie and his team also recovered parts of skeletons of two duckbill dinosaurs. 'One idea that was floating around was that maybe the duckbill dinosaurs were the reason the albertosaurs were all there,' Currie says. 'They killed it or perhaps they were scavenging. The only problem with this idea is that we don't have any tooth marks on the duckbill bones. So I think the duckbill dinosaurs are there coincidentally. They were brought by floodwaters and just deposited among all the *Albertosaurus* remains. However, it is unlikely the *Albertosaurus* bone came in by floodwaters because most of the articulated specimens are found in one place.'

All of the *Albertosaurus* bones were spread in one direction (north) in relation to Brown's 1910 quarry. There are no bones to the south, even though the layer continues for more than a kilometre. 'This is where they died and the bones spread downstream, to the north, as the bodies decomposed,' Currie concludes.

* * *

At the same time Currie and his team were excavating the *Albertosaurus* bone bed, more evidence came from Currie's colleague in Argentina, Rodolfo Coria. Currie and Coria had been working on a site in Patagonia in February 1997 that they originally thought contained a single skeleton. However, after some of the material was prepared in the lab, Coria sent Currie a fax saying that the site had at least three individuals of the same huge carnivorous dinosaur. Initially they thought the theropod was *Giganotosaurus carolinii*, an enormous species that had been described by Coria in 1995. It had a tremendously large skull measuring some 2 metres (6½ feet) and weighed as much as 13 tonnes.

Currie returned to the site in northern Patagonia the following year to collect more material. After five years of working the site, Currie and Coria had dug out the biggest fossil quarry in South America and uncovered evidence for nine individuals of an animal closely related to *Giganotosaurus*, which turned out to be a brand new species of animal known as *Mapusaurus rosei*.

'The evidence was very, very strong there that in fact we had a similar kind of situation to the *Albertosaurus* bone bed,' Currie says. 'These were large carnivorous dinosaurs that had died together and that probably were, in fact, living together up to their time of death. And in the case of the *Mapusaurus* bone bed in Argentina, the evidence was even stronger in that there were no other animals there to confound the situation. The site was composed completely of *Mapusaurus* bones, except for a few sauropod footprints where the animals had stepped on the *Mapusaurus* bones and pushed them into the mud.'

Since that time, Currie has been involved with several other sites in the north-west US state of Montana that have accumulations of tyrannosaur bones. One bone bed had three hadrosaurs and five specimens of the tyrannosaur *Daspletosaurus*. 'Again,' Currie says, 'that indicates that those animals were concentrating in that bone bed for a reason and that reason was probably associated with packing behaviour before their death.'

Currie also points out that the site where the *Tyrannosaurus rex* specimen known as 'Sue' (nicknamed after its finder, Sue Hendrickson), the largest and most complete *T. rex* specimen which was found in 1990, also provided evidence of pack behaviour. While the specimen was being prepared at the Black Hills Institute in Hill City, South Dakota, the scientists realized that 'Sue' wasn't alone in the quarry. There were the articulated hind legs of another dinosaur, and also isolated bones of a very small individual. Again, this was a site with a rare carnivorous dinosaur that contained at least three individuals.

'As time goes on, we're finding more and more sites like this worldwide. The evidence is saying loud and clear that carnivorous dinosaurs, just like the herbivores, are in fact animals that from time to time at least would pack together, probably to hunt.'

* * *

Currie is not afraid to entertain and study other theories for the *Albertosaurus* bone bed. These competing theories force him to re-examine the evidence and the strength of his own theory. 'When you're excavating a bone bed, you're looking at a crime scene that's 70 million years old in this case,' he says. 'There are a lot of clues that have gone cold on you and there are things that could give you one idea but in fact be indicating something that's totally different. You have to consider all these things when you're excavating a bone bed.'

David Eberth, who has worked in the *Albertosaurus* bone bed, does not believe that the site is conclusive evidence of pack or gang behaviour. As Currie has collected more and more specimens of only carnivorous dinosaurs from the

'These competing theories force Currie to re-examine the evidence and the strength of his own theory.'

locality, he and Eberth have debated his theories. Eberth has examined the site and, for starters, he believes that there are far more dinosaur specimens at the site than Currie has collected.

'There is absolutely no way just given the way that the landscapes work and erosion works and rocks are preserved [that] there aren't more animals in there,' Eberth maintains. 'There is no way that Phil was able to collect exactly the number that were preserved at that locality. The locality continues to go into the rock [and] is larger than what he has excavated, and there are parts of the locality that have eroded away. It is just inconceivable that the exact number that was preserved there has been pulled out of the rock.'

His reasoning stems from the fact that there is no analogous situation that has only carnivores present. 'There is no ecological situation that I know of where you could maintain a large group of very large meat-eating dinosaurs in a social structure on an on-going basis with those numbers,' Eberth says. 'There is no modern analogue for such a thing; there is just absolutely no evidence for it. However, there are lots of modern analogues and lots of examples of situations where large predators come together temporarily. They tolerate one another because food resources may be abundant or they may be threatened environmentally.'

Eberth points to grizzly bears as an example. If fish are in abundance in an area, grizzly bears will tolerate one another there. If something happened to those animals at that time, scientists would ask years later whether those animals lived in some kind of social structure or did they happen just to experience some kind of mass-mortality event. Based on what is known about the biology of grizzly bears, Eberth says the answer would be that it was some kind of unusual circumstance and did not reflect normal biological behaviour on the bears' part. 'It's far more reasonable to interpret the

Albertosaurus bone bed as [resulting from] some kind of environmental pressures that are drawing these animals together.'

This dovetails with the theory that maybe all these carnivores were very hungry or very thirsty and there happened to be a food or water source there that drew them all to the same area. There are several sites in the world from different ages that are known as 'predator traps'. What happens is that a plant-eating animal of some kind becomes trapped in quicksand or in tar sands, like those of the Californian La Brea Tar Pits. Its dying cries attract a carnivore. The carnivore moves in for a meal and becomes trapped. Then the carnivore's dying cries attract another carnivore or scavenger, and the next thing you know you find that for every one plant-eating animal, there are carnivores coming in time after time, and becoming trapped in succession. Predator traps do exist and they do work, and to this day there are still are predator traps that can be studied to see how they function.

Donald Henderson, the Royal Tyrrell Museum curator who disputes Currie's theory that tyrannosaurs were gregarious, does not doubt that they died together at the *Albertosaurus* bone bed. However, he doesn't believe that pack behaviour was the reason. He offers several theories for such a high number of *Albertosaurus* specimens being found in the same place.

'One argument is that they may have come together as separate carcasses drifting in and got

OVERLEAF
Grizzly bears will tolerate one other if there is enough food to go around. Did dinosaurs behave similarly?

jammed up,' he says. 'Or they may have died during a feeding frenzy. And it's been seen [with] Komodo dragons, there's often a lot of injuries and mortalities at these feeding sites. Adult Komodos will eat smaller Komodos if they're in the neighbourhood. It's an easy meal. And I suspect that some of that assemblage of bodies could have been related to this sort of intra-specific aggression between tyrannosaurs.'

Picking up on the predator-trap theory, Henderson speculates that if there were a large, smelly carcass that attracted the meat-eating dinosaurs in the first place, then it would draw a large number of them from far way. 'Tyrannosaurs were equally active predators, but they wouldn't give up an easy dead carcass as well. And I suspect a lot of these animals would have been brought there by a sense of smell. They may have even heard their own kind making noises at a site [which] drew them in.'

Currie has considered this possibility for *Albertosaurus* bone bed. 'It's possible that what Brown had found was a predator trap, but when we looked at the site in detail we realized that there were problems with that interpretation,' he says. 'First of all, all of the bones are at one level. There's no evidence of quicksand, and there's certainly no evidence of tar pits or anything like that. The bones all show the same kind of preservation, so there's no question that it's not a long-term accumulation. This is a very short-term accumulation of bones. We have no evidence of tooth marks on any of the bones, either on the predators or the one herbivore that we have in there. Therefore, there's no indication that they're going to eat something as either scavengers or predators.'

Drought is another environmental pressure that could have brought the carnivores together. Perhaps all these animals came to the same pool of water and, as the water dried out, they just died at that spot. These kinds of accumulations exist both in the fossil record and in the modern world as well. 'The trouble is that in those kinds of accumulations again you don't tend to find one group of animals, and you certainly don't find carnivores concentrations,' Currie points out. 'What you find generally are herbivores and you'll find many different species of them. Very often you'll find carnivores mixed in with them as well, because they are so thirsty they don't care any more about eating ... they're just there to get the water.'

Currie explains that it is very rare in a drought situation to find only one species of animal. Invariably, in those situations just about everything in the area is attracted to the remaining water. Based on this and other evidence from the bone bed, Currie is able to rule out that the bone bed was an accumulation caused by drought. 'We have no mud cracks in the bed, nor the drying and breaking up of the bones themselves,' he says. 'So it's another idea we can eliminate. Maybe there was something in the area that *Albertosaurus* liked and would come in to consume; but so far we haven't found any evidence of that.'

For Currie, the evidence all points in one direction. 'Over the years I've pretty much come to the inescapable conclusion that these animals were in the same place because they died together, and they died together because they were living together during at least their last few days of life. That suggests that there is some kind of gregarious behaviour that has to be accounted for.'

The centenary of the *Albertosaurus* bone bed's discovery was celebrated in August 2010. Currie organized one last dig there in the summer of 2010 and then moved on to other locations. He hopes that somebody will continue the work in the future. To ward off any possibility of poachers, Currie and his team do not uncover more bones than they can remove, so an amateur bone hunter would walk over the site and not spot a thing. However, to make things easier for the next legitimate dino hunter, Currie will list the coordinates of the site in the collections at the University of Alberta, at the American Museum of Natural History and at Alberta's own Royal Tyrrell Museum. Thus far, the site has produced hundreds of *Albertosaurus* bones, some of which are displayed in an exhibition just inside the entrance of the Royal Tyrrell Museum in Drumheller.

* * *

Currie knew that one or even two examples do not prove a theory, so he needed to examine another collection of tyrannosaur skeletons. To further his theory, he knew that he needed to spend time in a place that was home to *Tarbosaurus* and unlike anywhere else in dinosaur history: Mongolia.

'Once we knew that *Albertosaurus*, which is a smaller relative of *Tarbosaurus*, congregated in packs, then of course our eyes were opened up to the possibility that *Tarbosaurus* might also do the same thing,' Currie says. 'Of course, when you have two closely related animals, they can have very different behaviours. You can look at a lion versus a tiger for example, and you can see that they do behave in very different ways, with one packing and one not packing. So it's quite possible that in the case of *Tarbosaurus* and *Albertosaurus* they were doing very different things too. And yet, it's quite possible that *Tarbosaurus* was a packing animal as well.'

Currie would go to Mongolia with the idea that he would not necessarily find the same kind of thing he found in Alberta, but at least he would look for evidence of similar behaviour in *Tarbosaurus*. With *Tarbosaurus*, Currie would be dealing with a greater number of specimens, which meant that he had a very good opportunity to better understand what they were doing inside their ecosystem – and whether or not his dino gang theory would hold up.

chapter 4

the gobi desert

The bleached reddish-brown plains of the Gobi Desert stretch as far as the eye can see, meandering through long, broad basins banked by low mountain ridges. The plains are broken up by occasional patches of grass and dried-out streambeds. What roads there are consist of sand and gravel. Camels, sheep and goats roam at will, and the nomadic residents live in felt-insulated mobile homes to ward off the extreme temperature swings that send the mercury down to a subfreezing –40°C/°F in the winter and upwards of a scorching 50°C (120°F) in the summer. Only 10–15cm (4–6 inches) of rain fall the entire year. But for dino hunters, this is heaven.

The world's fifth largest desert, the Gobi occupies nearly 130 million hectares (500,000 square miles) across southern Mongolia and down into northern China. There are five different eco-regions of the Gobi, each with varying terrain and climates. In a relatively small corner located in the north-western part of the desert is one of the world's greatest treasure troves of dinosaur specimens, the Nemegt Basin.

'The two biggest sites in the world in terms of dinosaurs are Dinosaur Provincial Park in Alberta and Nemegt Basin in terms of numbers of skeletons recovered and in terms of the density of Late Cretaceous dinosaurs,' Currie says. 'The diversity of the sites is quite amazing, and they represent almost the same time period. The individual sites are roughly 70 to 75 million years old.'

In Mongolia, there is a greater range of diversity of the habitats because Mongolia was a much larger landmass than western North America was in the Late Cretaceous. Consequently, it was much dryer towards the interior and much wetter towards the outside, which gave a range of environments in between. 'In places you had big rivers running through dry, arid areas and it's in these corridors that some of these dinosaurs could live along and pretend that they had the ideal environment. It was just very restrictive in terms of area. We don't see quite the same thing in North America. That doesn't mean it didn't exist, it just means that we haven't found them at this stage. I think it is unlikely given the size of the landmass that we would have exactly the same environments as some of the ones we see in Mongolia.'

The Mongolian Gobi Desert was first explored for fossils by the American Roy Chapman Andrews in the 1920s. Andrews was looking for evidence of the origins of man, but instead he found dinosaurs, lots of them. He loaded tonnes of ancient dinosaur bones onto camels and carried them across the desert so they could be shipped back to the American Museum of Natural History in New York. Andrews also took along a documentary filmmaker who filmed hours of footage. The somewhat unexpected expedition became a touchstone event in the history of the exploration of dinosaurs.

'Roy Chapman Andrews certainly had a tremendous influence as the leader of the expeditions back in the 1920s, but also as a person who popularized science by writing books that influenced kids like me,' Currie explains.

Soon after Andrews' expedition, the area and its amazing dinosaur resources came under Communist control and were entirely closed off to Westerners. To most of the world, it remained a tantalizing, inaccessible mystery for almost 70 years. During that that time, Russian and Polish missions continued to search and excavate the area, and they found several different species of dinosaurs, among them the monstrous, meat-eating *Tarbosaurus*.

In 1964, eight *Tarbosaurus* skeletons were found in small area by a group of Mongolian palaeontologists. They remained exposed for many years because the site was recognized as being a significant dinosaur burial ground and the area was protected by the Mongolian government. Unfortunately, a few unscrupulous Mongolians discovered that dinosaur bones could make them money on the black market, so they found the site and hacked up all the specimens. At that time, teeth and claws were in the highest demand so most of the whole skeletons were destroyed in the careless extraction of these items.

To date, there have been more than a hundred confirmed *Tarbosaurus* finds, of which Currie has personally documented 90 sites in the field, up to and including the 2010 field season. A great number of *Tarbosaurus* have been excavated including those collected after the area was re-opened to Westerners in the late 1980s. They are now housed in museums around the world, including the Polish Academy of Sciences, the Palaeontological Institute of the Russian Academy of Sciences in Moscow and the Fukui Prefectural Dinosaur Museum in Katsuyama, Japan.

At least three more *Tarbosaurus* specimens have been found by Japanese expeditions in the Gobi Desert, but the quarries still have not been revealed. At some point, Currie hopes to work out a diplomatic solution so he can study the skeletons, but that will require a trip to Japan and some politicking.

'There's no question that the Gobi was *Tarbosaurus* central,' Currie says.

The fact that so many *Tarbosaurus* have been found in relatively close proximity to each other raises questions about the behaviour of these carnivorous dinosaurs. To date, there have only been approximately 35 *T. rex* found in North America, and they are scattered across different locations. Working in the Gobi, Currie has focused on two things to test his dino gang theory. First, he wanted to determine that the *Tarbosaurus* found there are embedded in the same levels of the rock, meaning that they are from the same time period.

'To date, there have been more than a hundred confirmed Tarbosaurus finds.'

The question is, did they all die close in time to each other? Secondly, even if they all died together, the bigger questions are, did they live together, and if so, how did they live together?

Jack Horner, for one, takes a sceptical approach to the Gobi finds. 'First of all, you have to demonstrate that there is good evidence to suggest that they died at the same time, and then even if they did die at the same time, what were they actually doing – were they on a walk together, were they hunting together, were they off to the grocery store?' Horner says. 'It's a real stretch to say what they were actually doing if they were just lying there dead. It's like walking about in the field and looking at five dead cows that are lying there. Do you know that they all died there at the same time? If you determine that they did, could you actually say what they were doing there?'

There are various reasons why the skeletons could have been fossilized together even if the animals didn't live together. The dinosaurs could have all been drawn to the river from great distances for water and then died in a catastrophic weather event. Or the bones could have been washed together by river currents after the dinosaurs died. As he explored his theory, Currie would have to consider every possibility and weigh the evidence before drawing any conclusions about behaviour. In the process, he would call on the expertise of other scientists as he refined his ideas and conducted his own new research.

Currie has been going to the Gobi Desert since the 1980s. His first full expedition was in 1986 as part of the Canada–China Dinosaur Project, the first cooperative palaeontological partnering between China and the West since the Central Asiatic Expeditions of the 1920s. Currie led the expedition on his home turf of Alberta, Dale Russell led the group in the Arctic and Dong Zhiming led the group in his native China. Serendipity played a role in putting together the expedition.

When Currie was setting up the Tyrrell Museum in 1981, he moved out of his offices at the Provincial Museum of Alberta because he didn't want it to appear that the Tyrrell was going to be a satellite of the parent museum. One of the people working for him was Brian Noble, who had been a naturalist in

Dinosaur Provincial Park and was designated the publicity manager. However, when the museum's development group moved to Drumheller, Noble decided that the town was too small for a permanent base and so he stayed in Edmonton. He asked Currie how else he could help.

'In the early 1980s, things were starting to open up in China and Mongolia, and I decided it would be pretty damn cool to go there,' Currie recalls. He had read about the Andrews' expeditions and the Polish and Russian finds, and he knew that he needed to work in the 'mother load of dinosaurs' to further his knowledge and expand his theories on dinosaur behaviour.

'For me, Central Asia wasn't so much about going to where my childhood dreams were in terms of following in the footsteps of Roy Chapman Andrews. It was because scientifically it was very interesting and potentially very productive,' he explains. 'I knew that in Alberta we had all these dinosaurs, but I knew that there was an emphasis on preservation and collection on the big guys. The big duckbilled dinosaurs and tyrannosaurs are pretty common as fossils in Canada.'

In Asia, it was the other way around, and at many sites it was the smaller dinosaurs that were more likely to be preserved. The region was home to many small specimens, such as *Velociraptor*. Palaeontologists knew from teeth and isolated bones that both North America and Asia were home to the same animals, but in Asia they were getting whole skeletons. Currie needed to see for himself what the differences were that led to these preservational biases.

'I had to make some kind of sense of what's going on in Central Asia versus what's going on in North America. At that point, the connection that I was mostly interested in trying to sort out [was] how close or how different are the faunas – how similar are they in terms of preservation – of Central Asia and of central Alberta.'

There were ideas floating around, but they were very controversial and conflicting with one another. Currie says that the American expeditions in the 1920s stated that some of these dinosaurs were living in what were basically sand dunes. Therefore, for example, the *Oviraptor* found on a nest of eggs by the 1923 Andrews' expedition was there because it had been caught in a sandstorm and the sandstorm buried the animal with its food. Andrews had concluded that these were dry-land environments. However, the Russian expeditions in the 1940s came to a different conclusion. Their

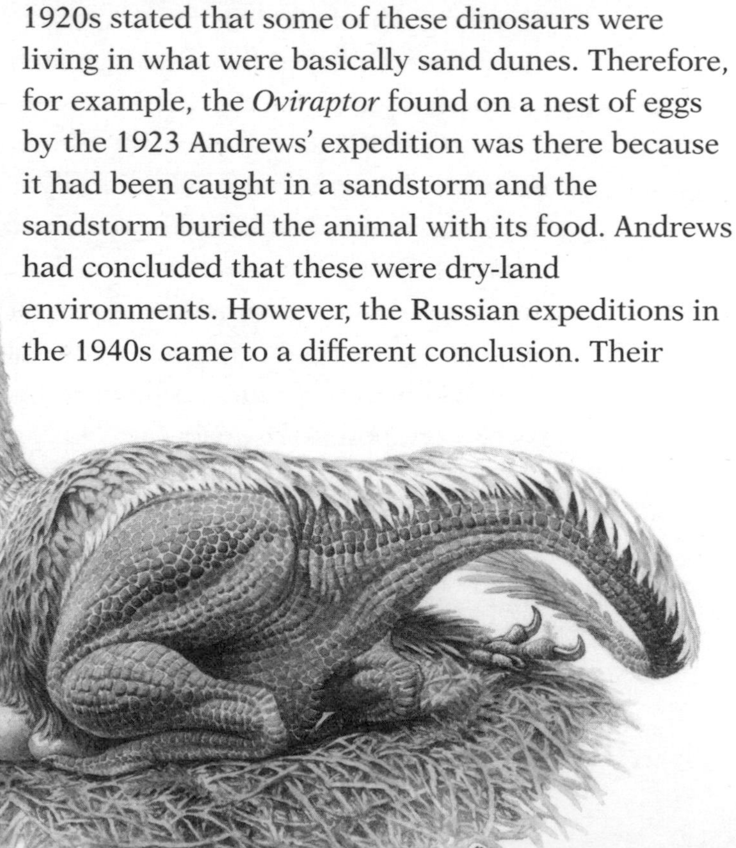

Oviraptor *brooding its nest of eggs. Specimens representing at least four species of theropod dinosaurs have now been found on nests of eggs.*

interpretation was that the area in the dinosaurs' era contained quicksands and wet environments. These were diametrically opposed interpretations, clearly based on old-school science, and they did not even consider the possibility of gregarious behaviour.

Brian Noble set to work investigating the possibility of sending Currie to Mongolia. An international political dance for access ensued. At one point, the Canadian Ambassador travelled to the Mongolian capital of Ulaanbaatar to lobby for Currie and his team to be allowed access to the Mongolian sites, but the talks ended in a stalemate. Currie then switched his attention to the area of the Gobi Desert that reaches into north-western China. Up until then, China had been even more closed to foreign palaeontologists than Mongolia due to its restrictive Communist government.

China was something of a big unknown in terms of fossil finds at that time because few dinosaurs had been collected there that were similar to the ones in Mongolia. Most of the material that had reportedly been found was in pretty bad shape, such as chunky bits of *Velociraptor* or *Protocerotops*. For Currie, there wasn't a lot to go on, and yet logic told him that northern China had the same kind of beds as there were elsewhere in the Gobi, and therefore the dinosaurs should be there somewhere.

The Chinese were being wooed by several institutions for access at this time, but the Tyrrell Museum was already making headlines and Chinese officials were aware of the Alberta resources, thanks mostly to Currie's recent finds. There was also an unusual connection between Canada and the Chinese in Norman Bethune, a Canadian doctor who had worked with Chairman Mao and had become one of the most famous people in China. As part of the bargaining, the Canadians invited the Chinese palaeontologists to come to Alberta and hunt for dinosaurs with Currie's aid.

After a year-long back-and-forth over the logistics, an agreement was put in place. Interestingly, all the publicity in Canada was about the expedition opening doors to China, whereas the Chinese publicity was focused on their scientists going to Alberta rather than about China opening itself up to Westerners. The final signing ceremony was held at the Tyrrell Museum only months after the museum opened its doors to the public in 1985.

In 1986 Currie and his fellow dino hunters started their trek in north-western China and then worked their way across the Gobi Desert. There was one particular site that resembled the Flaming Cliffs of Mongolia, where many important dinosaur finds had been made by the Andrews expeditions in the 1920s. But the difference was that the new expeditions were finding some of the first dinosaurs from the Chinese Gobi. In one site, they made a landmark discovery: 12 baby ankylosaurs, or armoured dinosaurs, had died upright in a sandstorm.

'Probably what they were doing was lying behind a sand dune,' Currie theorizes. 'They were trying to escape the blasts of sand in the eyes. If you go behind a dune, you are shielded from the ferocity of the wind but, of course, the sand is still coming over the top of the dune raining down on you. I think what happened is they got buried too deep and they couldn't get out. That was the first indication of gregarious behaviour in the Gobi. We also found a *Protoceratops* site that was showing the same thing.'

Over the course of the five years of expeditions, the group collected over 60 tonnes of material. 'It was unbelievable,' Currie recalls. 'Every single day was a new discovery.'

After returning to China over the following three years, Currie finally went to Mongolia in 1989 and travelled to the Flaming Cliffs. He had hoped to meet with Rinchen Barsbold, the top Mongolian palaeontologist who is an expert on theropod dinosaurs, but Barsbold was trapped in a different part of the desert by vehicle breakdowns. Barsbold was well known around the world because he had worked on the Polish-Mongolian and Polish–Russian expeditions and had published several key scientific papers on Central Asian dinosaurs. Currie eventually did meet him and was able to question him about the Polish and Russian expeditions, and in recent years has worked with him as part of the Korea–Mongolia International Dinosaur Project.

Over the ensuing years during his regular trips to the Gobi, Currie spent his time attempting to incorporate his finds from Canada's badlands – notably his rediscovery of the *Albertosaurus* bone bed – with what he discovered in the sites he visited in the Gobi. From the trackway sites Currie documented in Peace River Canyon, it was clear that big herbivores were not only moving in groups but actually bumping into each other. This meant that the idea of carnivorous dinosaurs living in groups was less far-fetched than previously thought. But there was still a lot of evidence that needed to be found.

One of the first things Currie had to contend with was that there were differences between the Alberta carnivorous dinosaur finds and those in the Gobi, and in particular the conditions and processes by which the dinosaurs fossilized. The environments in Alberta and the Gobi were different at the time that death occurred, which led to different preservations.

'These are big river deposits, but the problem is that the beds in Mongolia are working in a different way than the beds in Alberta,' Currie says. 'The Alberta beds were close to the sea and you were getting a massive amount of sediment coming in all the time carried by rivers from the uplands to the west, whereas at many sites in Mongolia it probably has a lot to do with windborne sand. Basically, what happens is that when things get buried in Mongolia, the whole animal is getting buried. In Alberta, we can get that kind of preservation too, but then we also get these long periods when animals are dying and their bones are accumulating. They eventually get picked up by the rivers, and buried in sand bars. This led to the creation of bone beds, which we don't have in the Gobi.'

Though there are an abundance of *Tarbosaurus* footprints in the Gobi Desert, there are no *Tarbosaurus* trackways – or at least none have been found. 'The trackways have to be there but the big problem is that because the beds erode into cliffs, the tracks come out of the cliffs one by one,' Currie explains. 'The trackways have to be there. It's just a matter of time and enough people looking.

'There are an abundance of* Tarbosaurus *footprints in the Gobi Desert, but no trackways.'

Eventually we are going to find a site where there are good trackways from *Tarbosaurus*. We do have older beds below the *Tarbosaurus* beds where there are good trackways, but then there are no skeletons and they have nothing to do with *Tarbosaurus*.'

Currie compiled the GPS coordinates of 66 *Tarbosaurus* finds in the Gobi Desert, and over a number of years he began visiting the sites to study the skeletons or peruse the sites themselves. Despite the removal of some skeletons and the fact that a few quarries had been poached, the remnants of many *Tarbosaurus* were still embedded in the rock. For Currie, visiting these sights was as close as he could come to being there and observing the tarbosaurs. But the sites continually raised questions about whether the dinosaurs had died and lived together – and more importantly, how they lived and died.

By 2006, Currie was anxious to return to Mongolia to examine all the *Tarbosaurus* sites and put this pack theory to the test. As with all expeditions, this would require money. That year, Yuong-Nam Lee, a long-time friend of Currie's and head of the Korea Institute of Geoscience & Mineral Resources, was able to secure financing for a five-year expedition from Korea's Hwaseong Dinosaur Lab, a cutting-edge facility for examining dinosaur egg finds in that part of the world. The Korea–Mongolia International Dinosaur Project would be led by Yuong-Nam and Currie. The other scientists who joined them on the expedition, including Louis Jacobs and David Eberth, would be partly financed by their universities and, in addition to helping Currie, they would all be looking for different things to further their own research.

For palaeontologists, the conditions in the Gobi veer between challenging and treacherous. Rain and sandstorms are the order of the day, but they don't occur in a normal pattern. Some days, a sandstorm will blow through and stop work for hours. Even worse, when it rains, sandstorms usually follow. 'Those days are kind of miserable,' Currie says. 'You just kind of hunker down and you try and survive – though even in your tent you can't escape from the sand.'

The team trucked in freezers and refrigerators and generators to power them. But every once in a while, generators will break down or the refrigerators will stop working. Something always goes wrong. On those days, the support staff travelled to one of the small villages or to a yurt camp to make a deal for some fresh meat.

Creature comforts aside, Currie calls the Gobi one of the best places in the world to hunt dinosaurs. 'It's a wonderful environment in its own right,' he says. 'It's very sterile, and it's a place you can go and be healthy almost all the time. Once you're used to the heat, it doesn't stop you from doing any work. There is no day when it's continuously raining, so you can work all the time. It's certainly not everybody's cup of tea, but if you're used to the desert it is a good place to work.'

The Gobi's barren desert floor is ideal for spotting dinosaur fossils. 'Looking for fossils has one basic rule to it – you need to look down,' Louis Jacobs explains. 'If you come to the Gobi where there's no vegetation, you can see the rocks. The rocks have to be sedimentary rocks that were formed in ancient streams and ponds and lakes and rivers or sand dunes, anything like that where you can imagine that the particles of sediments would cover up the carcasses and bodies of the animals that die. And then those sediments turned into stone, and you have the fossils.'

The present-day Gobi is no longer forming sediments that are burying things because its surface is constantly eroding. From the time that the rocks in the Gobi were formed, erosion has taken place and caused the bones of those animals that lived at the time the rocks were being formed to be worn away. When bones are spotted in the sediments, palaeontologists must follow the trail of fragments to the layer where they are coming out and then dig to see what has been preserved. The rewards can be great because many of the trails lead to nearly complete skeletons.

Despite the fact that the Gobi is so plentiful with dinosaur bones, palaeontologists have to be very selective because some specimens are very badly eroded. Though many specimens have been excavated over the years, there is always a chance of finding more, and in spite of the fact that the area has been intensively worked since the 1940s the promise of finding something new is still greater than in other locales. 'The reason that the Gobi is so special,' Jacobs says, 'is because people have looked in the Gobi for a long time and the Gobi continues to produce secrets over and over and over again.'

Seventy million years ago the Gobi Desert was a very different place from what it is today. Where we now see sand and rock, the Nemegt geological formation was once a river system with channels up to 30 metres (100 feet) wide. The hot and dry desert was then composed of marshlands, forests and plains with lush plant communities. The land was filled with herds of duckbill dinosaurs, and the skies were filled with flying reptiles, primitive birds and insects. The rivers, lakes and wetlands teamed with wildlife, including turtles, crocodiles and fish. Currie estimates that the Nemegt site in the Gobi has as many as 40 species of dinosaurs. There were small meat-eaters such as *Velociraptor*, and plant-eaters such as the hadrosaur *Saurolophus*. At the top of the food chain was *Tarbosaurus*, terrorizing all, eating anything in its path.

'This was an area that was obviously very good for dinosaurs to live [in] because we had absolutely huge animals living there,' Currie says. 'That included things like sauropods, the long-neck dinosaurs, an animal called *Opisthocoelcaudia*, giant *Saurolophus* and, of course, *Tarbosaurus* as

A reconstruction of the Cretaceous landscape.

the hunter or the predator in this particular region. What we're looking at now is a pretty dry desert environment, but 70 million years ago it was obviously a much more lush environment and a much better place for dinosaurs to live.'

In the Late Cretaceous period, the Gobi was what scientists call a mesic environment, or one that has a balanced supply of moisture throughout the year. According to sedimentologist Dave Eberth, an area that bears a striking resemblance to what the ancient parts of southern Mongolia would have been like 70 million years ago is Pantanal in Brazil, which is the world's largest wetland.

'This was an area of seasonal wetlands,' Eberth says. 'It went from being dry to being very wet on an annual basis so during one part of the season this area is a very lush and wet area with lots of aquatic animals moving around in it, and in another part of the year it gets quite dry. During that dry season, we find more terrestrial animals moving into the area and exploiting the resources. So think of it as an extensive marsh area with little meandering streams running through it and also experiencing these annual changes from being warm and wet to being warm and much drier.'

Eberth has re-created what the Gobi would have been like at the time the dinosaurs lived there. He documented in detail the environments, from the size of the rivers and how fast they flowed to the frequency of the flooding events, how much water was here and what that climate was like. 'That kind of detail is absolutely phenomenal to retrieve from an area as remote as this, and it's going to give us a picture of life at the end of the Cretaceous,' Eberth says. 'We're going to be able to make comparisons now between this part of Mongolia, other parts of Mongolia, northern China, North America, Europe, and so on and so forth.'

Currie explains the pecking order 70 million years ago. 'There's no question that *Tarbosaurus* was the top predator in this region, and it shared the role with many other predators, the same way as lions share the role with cheetahs and hyenas and animals like those,' he says. 'But nevertheless, there is no question as to who was king here, it was *Tarbosaurus*. *Tarbosaurus* was definitely the most sophisticated predator in this region, and was also the largest by a long shot. So it was the animal that all the other ones had to move out of the way of, shall we say.'

The Gobi is the ideal place for Currie to ask critical questions about gang behaviour. If *Tarbosaurus* was the apex predator, why are there so many concentrated in a relatively small space? Typically, apex predators need large areas to themselves to have enough food. So if they did live together, how did so many meat-eating dinosaurs co-exist in the same area? They would have had to cooperate in some way and not battle one another. Is it possible that they also hunted cooperatively? As Louis Jacobs puts it, 'If there are [all those] tarbosaur specimens, that's a hell of a lot of hunts, a

TOP
A reconstruction of a Tarbosaur kill.

BOTTOM
Wolves hunting cooperatively: modern-day successors to dinosaurs?

hell of a lot of kills, and a hell of a lot of mouths to feed.'

One of the best sites in the Nemegt Basin is called Bugin Tsav. It contains at least 20 *Tarbosaurus* skeletons. Currie enthuses that you can stand in one place and within your field of vision you can see six quarries where *Tarbosaurus* skeletons have been excavated in the past few years. Currie focused on this site because, in spite of the fact that the animals are separated from each other, there is very strong evidence that the animals died at approximately the same time, if not at exactly the same time.

'So far what we've discovered is that there's no question that several of those half dozen or more specimens definitely did die and get buried at the same time, and we're talking within days of each other here, possibly even within hours of each other,' Currie says. 'But with some of the other ones, because they're separated by a slightly larger distance, it's a little harder to interpret them because in a river system you have a lot of sediment being moved and reworked all the time by river waters. It makes it very difficult sometimes to be 100 per cent sure that those animals did die at exactly the same time.'

Eberth theorizes that a large variety of dinosaurs lived along the banks of those rivers and also lived in the areas between the rivers in the lush, lowland area. He believes this may account for why so many *Tarbosaurus* skeletons are found near each other.

'Every once in a while there would be a major flooding event coming in, and there's no high ground here,' he explains. 'You have a big flooding event coming through the area, and that's going to kill animals, it's going to drown animals, plus it's going to pick up the carcasses of animals that were already dead and rework them into the channels.'

The different sites do not have layers of rock that conform to one another. Therefore, even though 17 tarbosaurs have been found in the same level of rock (suggesting that they all lived at the same time), because of the way the sediments are, Currie and his team cannot be certain the different groups are all related.

'That structure in the rocks has a three-dimensional shape,' Currie explains. 'So if you find the specimens at different levels, and your beds are sloping in a certain way, unless you can actually connect the dots in between the two sites you can't be 100 per cent sure that the three animals found in one part of the sandbar are from the same time as the three animals found in another part of the same sandbar. So all of those animals found at Bugin Tsav – a dozen or more tarbosaurs – may in fact be animals that died at exactly the same time in the same event. But the trouble is, we can't prove that more than three or four of those animals did die together at the same time.'

One of the reasons that Currie works with geologists like Eberth is to make sure he is reading the rocks correctly. The rocks in the Gobi are Late

Cretaceous. Not only are there dinosaurs in them, geologists have dated the rocks to the era with certainty using radiometric dating techniques (which are similar to radiocarbon dating methods used by archaeologists). But natural forces around the Earth over millions of years, such as ice ages, floods, earthquakes and volcanic eruptions, have actually brought together layers of the Earth that are thousands of years apart.

'The problem with the Mongolian bone beds is that we have no real way of dating them,' Currie says. 'It is a desert and the ideal thing you need for dating these kinds of beds is volcanic ash, lava, molten rock, or anything coming out of a volcano. As soon as you've got that, then you can radiometrically date the beds.'

Eberth sees himself as someone whose job is to constrain the biological hypotheses, including aspects of Currie's dino gang theory. 'If the palaeontologists were finding lots of dinosaurs and they were thinking, "Hey, we're finding lots of these carnivorous dinosaurs, *Tarbosaurus* or *Gallimimus* [an ostrich-like dinosaur]", they may actually start thinking that perhaps all these animals were living together at the same time and died together. Geologists like myself come in and we can actually start looking at exactly the right horizons that these animals occur in,' Eberth explains.

Eberth's primary task is to study the different layers of the rock strata to determine if the dinosaurs died together. He believes that the differences in the rock layers indicate that the specimens lived through thousands of years of geological time. By studying them closely over the years, Eberth has been able to show that a lot of these fossil occurrences that are laid out over this landscape are actually in subtly different layers of the rock strata.

'This means those animals didn't live and die together at the same time,' he says. 'We're being fooled in a sense because it's a flat area, and we're seeing the subtlety of the stratigraphy concentrating a lot of animals in an apparent surface. But there's actually a lot of differences in those layers, and we're probably seeing animals living through 500,000 years of geologic time [in this area].'

However, Eberth is cautious not to go too far with his assessment. 'The package of the strata is only 20 metres [65 feet] thick,' he continues. 'So, because of that I'm not going to say that the animals necessarily didn't all live together and didn't live in groups together. But we can certainly say that there is very little evidence from Bugin Tsav that they did. And we're going to have to find – if we're going to push our hypothesis that these animals are all living together, that they are communal if you will, that they live in packs or they live in herds – better evidence than is available from Bugin Tsav.'

Currie disagrees and believes that the evidence in Bugin Tsav is fairly strong.

'What we did see is that large numbers of *Tarbosaurus* were being found at the same levels in the Nemegt Formation,' Currie says. 'Because they're at the same level in the stream channel [deposits], then one can't help but think that they died at around the same time, in roughly the same place.'

Because there are such a large number of *Tarbosaurus* skeletons, Currie concedes that there is no question they did not die all at the same time. 'These were animals that lived for a long period of time, over several million years and they died at different times,' he says. 'And yet within that assemblage of more than 60 *Tarbosaurus* skeletons, we clearly have concentrations of them at certain levels. So now we start looking for other kinds of clues.'

One clue is that the Japanese expeditions found several *Tarbosaurus* skeletons in exactly the same quarry. Though they were mixed up together, they were articulated skeletons. 'That's a pretty strong statement that these are animals that almost certainly died together because *Tarbosaurus*, in spite of the fact that it's fairly common in Mongolia, still is a carnivore, and within an ecosystem it's not going to be an overly common animal,' Currie insists. 'So if you find several skeletons together in one place, then the option is there that we may find other sites where we have multiple specimens.'

Currie has found some evidence to support his dino gang theory. Instead of concentrating only on skeletons being found together, he has stepped back and looked at the big picture. 'The fact that we have so many *Tarbosaurus* skeletons [in the Nemegt Basin] suggests that something's going on that's a little bit different than Alberta and a little bit different than what we see in the majority of ecosystems,' he says.

In any natural environment, a top predator like *Tarbosaurus* living in groups would not be common, because it takes such a large amount of food to

'What is just a bunch of rock to most people, to us has so much information.'

support that top predator. 'If there were too many of them, of course, they would eat themselves out of house and home and they would die,' Currie explains.

Under most natural conditions, the top predator only makes up about 5 per cent or so of the fauna. Therefore in the Gobi region there should be approximately 19 skeletons of hadrosaurs and sauropods for every *Tarbosaurus* skeleton. However, that is not the case.

'We're finding *Tarbosaurus* makes up as much as 50 per cent of the animals at this particular level,' Currie says. 'Something's going on here and we really don't understand it. It's probably related to the behaviour of the animal in part, but there have to be natural circumstances for concentrating the animals like that as well. So what is just a bunch of rock to most people, to us has so much information, not only on the environment that dinosaur lived in, but also on the behaviour of that dinosaur.'

For Currie, this concentration of *Tarbosaurus* finds could be additional proof that they were cooperating, though he will need more evidence to solidify this idea. The conundrum means another trip back to the Gobi Desert to further examine all the evidence and to dig for more.

chapter 5

a dinosaur dig

Palaeontology is a rugged trade, and fieldwork is undoubtedly the toughest part. But this is where it all begins. Despite the advances in modern technology, most fieldwork is still done the old-fashioned way. Palaeontologists sleep in tents (eat whatever can be frozen or killed within a short distance from camp), bathe in rivers, and dig for dinosaur bones with shovels and picks.

A dinosaur excavation is, in turns, gruelling, frustrating and exhilarating. The palaeontologists live a fairly primitive existence and endure unusually harsh conditions no matter where they go. Depending on the time of year, there can be temperature swings between day and night of 25°C (77°F). Obstacles are provided by nature in the form of sandstorms, ticks, mosquitoes and wild animals, and by modern technology in the form of broken-down trucks and malfunctioning refrigerators. The big trucks have six-wheel drive to prevent them from getting stuck in the sand. They are built very simply and are easy to fix – provided you have the parts.

Hauling tools and the plaster for jacketing the fossils is a hard chore. Trucks carry them across dirt 'roads' but, once the palaeontologists reach a dig site, they must hand carry everything up and down the hills. Occasionally, when funding allows, helicopters are used to bring in compressors and plaster to preserve the fossils and then to airlift the plaster-jacketed fossils from the field.

In the last 25 years, Phil Currie has done fieldwork in Alberta, China, the United States, Mongolia, Argentina, Madagascar, the Arctic and Antarctica. Enduring the conditions is half the battle. One of the most gruelling places to work is Antarctica. Currie was there in 2003 and 2010. He has been working with a team to collect a single skeleton of a large theropod named *Cryolophosaurus*, which is embedded in very hard rock on Mount Kirkpatrick. The dig site is almost 3665 metres (12,000 feet) above sea level. Besides the reduced oxygen level that makes breathing difficult and causes premature fatigue, the temperature variance ranges from a high of –20°C (–4°F) to a low of –30°C (–22°F). 'It's a spectacular place to work,' says Currie, ever the rugged palaeontologist. 'I've never seen anything like it.'

The entire group was required to complete survival training to cope with the cold and ice and the altitude. The training teaches the palaeontologists how to hang on if they fall into a

Setting up the base camp for a dinosaur dig in Bugin Tsav in the Gobi Desert in 2009.

crevice or get caught in a snowstorm until help arrives. In one exercise, everybody was given an ice-cream bucket, which they had to place on their heads. They were then roped together and taken a short distance from their camp and told to find their way back. There were traces of light coming in through the sides of the bucket, but they basically couldn't see anything so the situation simulated a whiteout snowstorm.

Currie says that in 2003, only two of the five teams made it back. 'The greatest education was seeing how people reacted to leadership,' Currie says. 'When you have a lot of independent thinkers, they don't necessarily accept that somebody else is going to lead. There are four or five people who would stop dead and just argue about it – in the bone-chilling cold with an ice-cream bucket on your head!'

On each trip, Currie stayed in Antarctica for seven weeks, including the training period. After about three or four days, Currie says the body adjusts to the extreme cold and the 24 hours of sunshine. The group slept in sleeping bags in tents with no heat source other than their own bodies and their clothes stuffed into their bags for added insulation. In the mornings, they had to take their boots out from under their beds and break the ice out of the insides caused by the freezing of the sweat from the previous day. 'It took me 20 minutes in the morning to get dressed,' Currie recalls.

Currie relates one particular story. For years he received letters from an enthusiastic young boy who wanted to be a palaeontologist. The boy attended a seminar in his hometown of Ottawa given by Currie about fieldwork. After the talk, the boy approached Currie with a troubled look on his face and asked, 'Do you have to sleep in a tent all the time?' Currie replied that was the way it was. 'I hate camping!' was the boy's response, and Currie never heard from him again. 'You have to have that gypsy gene,' Currie says of his trade.

Currie and Koppelhus camped in the CTAM (Central Transantarctic Mountains) for six weeks during the 2010–2011 expedition to Antarctica. The mountain in the background to the left is Mount Kirkpatrick, which is 600 kilometres from the South Pole.

In fact, Currie prefers living somewhat primitively in the field. When Dinosaur Provincial Park was named a UNESCO World Heritage Site, the park officials upgraded the accommodation and built a trailer park for the dinosaur hunters to smarten up the place, and they passed a no camping rule. Currie hated it. 'The walls were so thin that you could hear anybody snoring three rooms away,'

he says. 'I prefer camping. There is more privacy. You can pitch your tent where you want to be and sleep in your own space. There are no lights when you are camping next to the Red Deer River, so you can see an amazing vista of stars. You have the prairies, the badlands, the river and the forest alongside the river – all within walking distance. And all of the different species of birds that live in these different environments are flying around your camp.'

In the summer of 2009, Currie returned to Bugin Tsav, the Gobi Desert site that has the greatest concentration of *Tarbosaurus* skeletons, with a group of expert palaeontologists for a dinosaur dig to further explore his dino gang theory. Joining Currie and Eva Koppelhus in the fourth summer of the five-year expedition were Yuong-Nam Lee, sedimentologist Dave Eberth, Yoshi Kobayashi, Japan's top expert in carnivorous dinosaurs, and the renowned vertebrae fossil expert Louis Jacobs.

Yuong-Nam works primarily on plant-eating dinosaurs, as well as on eggs and footprints, and he is also known as an expert fossil spotter. Yoshi is focused on ornithomimids, known as ostrich-mimic dinosaurs because of their resemblance to the modern-day ostrich. Currie, of course, is zeroed in on *Tarbosaurus*. In one single sediment layer in the Gobi, there are all of those types of dinosaurs. Together with Eberth and Jacobs, they sought to answer what kind of environment these animals are found in, what kind of environments they lived in, and what kind of environments they died and are preserved in.

Unlike previous digs, there was added pressure because Currie and his colleagues were being filmed by Atlantic Productions for a documentary that would air on the Discovery Channel in the US. Such programmes provide a larger showcase for Currie's work and can dramatically increase public awareness in dinosaur hunting, which can be instrumental in leading to funding for future expeditions.

'The beautiful thing about having these multidisciplinary teams is that when we all get together in the evening, we're asking questions,' Eberth says. 'We're all challenging each another. We're all looking at things slightly differently, for example the way that Phil and I look at meat-eating dinosaurs. This forces each of the members of the scientific team to make sure they collect enough data and to make their case. Everything that we do on these projects requires very, very careful thought and collecting enough data to make your case, and one of the best ways to push yourself is to have a team of other scientists around you, challenging your assumptions and challenging your conclusions.'

The group of palaeontologists and their local support staff lived and worked together in a place that is literally miles from nowhere. Food, water and everything they needed to survive was brought in and kept at the base camp. The food was prepared by the local help. When possible, they buy a goat from any local Mongolians who happen to be in the area. It's fresh the first day, but on the second day it can start to smell. To add flavour and disguise the taste, Currie brings 'Montreal Steak Spice' and Tabasco sauce with him. His wife takes it all in her stride. 'It's not 100 per cent Mongolian cooking but it's not bad,' Koppelhus says. 'Mongolians survive on it. Sometimes you have to live like the people you are with.'

The scorched sands of the Gobi Desert give the feeling of being on the surface of the moon. It is one of the harshest places on Earth – barren, isolated, inhospitable. The small dome tents where everyone sleeps offered little protection again the sandstorms that can arrive in minutes and last for days. Indeed, when the tents are uprooted by the winds, they become a liability. There are no mobile phones, no showers and nowhere to wash your clothes. To work out here, you've got to love what you do.

'There's lots and lots of people that want to be palaeontologists,' Louis Jacobs says. 'Every kid in the world wants to be a palaeontologist. But if you come to a place like the Gobi and live in a tent and your tent blows over in a sandstorm, and your food is full is sand, and it's hot, and you work for hours and hours and hours and days and days on something, and you get down to the wire and there may be one piece missing or something that you really wanted to get, or if your supplies run low or if the truck gets stuck, that's all part of it. Some people are not cut out for the discomfort and the troublesomeness of finding and digging fossils, and other people are better suited for it.'

Dave Eberth feels that the rough conditions concentrate the mind. 'One of the things that we often ask ourselves and other people ask us is "What's the attraction for being in such an extreme environment?"' he says. 'This is a really hot place, the flies are terrible, the resource support required to keep a large crew on the go can be quite trying for our Mongolian support staff. But the thing is that there's something about an environment like this where you're being challenged, you're focusing on learning. We're coming at this from a scientific point of view. The environment around us is a challenge, [but] we have to get past that. In some strange way, it kind of focuses your attention and makes you even all the more determined, regardless of the sandstorms and the flies and the gear failures, as we call it when things break down out here, which they do a lot. In spite of all that, we're determined to get the answers that we came to get. And that challenge attracts us to this area.'

Ticks are Currie's worst fear, and they are abundant wherever camels or other livestock have been in the Gobi. When they are engorged, they grow the size of a two-pence piece and their body expands out beyond their legs. All that's left is their hands and feet, and they basically drag themselves through the sand by their hands because they can't use their legs. He has nightmares about one encounter with a sea of ticks.

'We were stopped out in the middle of the desert with a mechanical problem,' he recalls. 'Whenever these things happen, for guys it's okay because you can go to the bathroom. It's not that easy for the girls, because you are out in the middle of nowhere and the landscape is bald and flat. We saw a corral in the distance. It was probably about a half mile away, so we walked to it. As we get there, Eva went around the back of the corral. Then I looked at the ground and saw that the ground was moving, and there was this wave of ticks coming out of the corral – literally, a wave. By the time I reacted, this wave was already coming up my legs. I ran back from the wave, knocked all the ticks off, and we got the hell out of there as fast as we could.'

After a few weeks, provisions run low at base camp. Coffee, burlap, beer and toilet paper are in particularly short supply. There is a debate going around camp about which is most critical. Coffee seems to be the least important. If the burlap (a type of sacking used to hold the plaster in place when the fossil jackets are made) runs out, old T-shirts can be substituted. Toilet paper is necessary for wrapping fossils before they are covered in plaster, as well as for its obvious use. In fact, the group agrees to ration it to ensure there will be enough – to pad the fossils, that is. As Currie points out, 'It's not like you can't find a big leaf.'

A consensus emerges on the most precious of the dwindling resources: beer. 'It's so hot and one of the best parts of the day when you're on an expedition is to get back to camp, have a beer, sit around with your colleagues, find out how the day went for everybody and watch the sun go down,' Jacobs says.

Day after day, the scientists rise early in the morning and head out in different directions to search for dinosaurs. It's a hard slog along what may be the world's toughest 'roads'. Trucks get stuck in the sand, heads are banged and exhaustion quickly sets in. When the trucks can't drive any further, the palaeontologists hop out and continue on foot. Across the vast landscape, the sun is searing and the wind is lashing and the combination makes it feel like your skin is peeling off. Littering the desert floor are the sun-bleached skeletons of camels and sheep that became separated from their herds, letting everyone know what can happen if you lose your way.

Currie has mapped out the entire area using a combination of old maps, a GPS and a computer. It is a complicated process. He travels with a file of the old photos of the Polish–Mongolian expeditions. On the map, Currie has marked their camps with red numbers. The Polish scientists set up cairns on the tops of hills and used their compasses to triangulate the locations to produce maps of the area. This is significant because it shows Currie the quarries that were excavated. Theoretically, if Currie wants to re-find those quarries, the ideal way would be to take an old map and superimpose it on a modern map. However, the modern maps aren't accurate either, and they are not as detailed as the Polish maps. Even the satellite imagery isn't adequate. So Currie has superimposed the old maps onto his modern-day map and then programmed his computer to distort the image and synchronize the coordinates.

Each day, Currie scours the barren landscape looking for more *Tarbosaurus* remains. Often, he spots something promising, only to find it's just a lump of stone. This process does yield some rewards, but each time it's a different dinosaur fossil – a duckbill or a small omnivore.

Three decades in the field have taught Currie a thing or two about tell-tale signs, and he will suddenly drop to his knees. A nodule of the red sandstone has caught his eye. He cracks it with his pick ... nothing. But as he's kneeling down there his eyes are darting back and forth, forth and back.

'In spite of the fact that it's been intensively worked since the 1960s, you can still wander around here and expect that you're going to find something new,' he says. 'And so we're always on the lookout for the exceptional while we're looking for dinosaurs here.'

At one point, Currie spots a large chunk of bone that he immediately determines by its spongy internal texture is from a *Saurolophus*, a duckbilled herbivore. *Saurolophus* were a little unusual among the duckbill dinosaurs because, instead of having a flat-topped skull or a hollow crest on its head, it had a huge spike coming out of the top of the skull. Currie believes the skeleton was probably broken up by somebody trying to find the skull of the animal for its sale value, rather than by any act of nature. Because the texture of a *Saurolophus* bone is very different form a *Tarbosaurus* bone, Currie doesn't

'Three decades in the field have taught Currie a thing or two about telltale signs.'

need a whole skeleton to be able to identify it. Of course, for a positive identification, down to species level, he will need better material than a single chunk of bone.

'Even a chunk of bone can tell us something about what may be under the ground here. If I was doing all my research on *Saurolophus*, I might come back here, pick up all these pieces, put them to one side, clean the surface off and just see where it came from. Then I would almost certainly get confirmation that it is *Saurolophus*, as opposed to one of the other duckbill dinosaurs. Experience means you need less and less to be able to identify things. When it comes to *Tarbosaurus*, a piece of a tooth is all I need to be able to determine it is a *Tarbosaurus* as opposed to anything else.'

Currie theorizes that the *Saurolophus* was probably one of the main prey items for *Tarbosaurus*.

RIGHT
Locals selling 'Gobi stones' alongside a road leading into the desert.

BELOW
A reconstruction of an environment with sauropod dinosaurs.

'Even though we don't find as many *Saurolophus* skeletons as we find *Tarbosaurus* skeletons, when we look at the footprint sites in this area there's no question at all, *Saurolophus* was the dominant dinosaur numerically. So it would have been the best animal for *Tarbosaurus* to go after to eat.'

Digging up dinosaur bones is an intricate process. One afternoon, a piece of *Tarbosaurus* leg bone is found by Yuong-Nam Lee, and Currie immediately makes his way to the spot before the excavation begins. When he arrives, he scratches around the site and more and more pieces surface. All the years of experience have taught Currie that there's one important thing to do when you find a piece of bone that nobody else has ever seen before: look up. The bones at foot level have usually eroded from higher up a rock face. Sure enough, that's the case here. There's more of this monster 20 metres (65 feet) up the cliff face, and it will take several people to remove the fossilized beast.

Before any digging starts, Currie inputs the location of the find into his GPS. As he does so, he realizes that other tarbosaurs have been found in close proximity. He studies his list of previous finds and soon sees that no fewer than seven other *Tarbosaurus* have been found in the area. Could this be the bone bed that he had been hoping for?

The process of excavation starts. The rule of thumb to digging is that the further one is away

TOP LEFT
Philip Currie explaining the process of plastering a fossil.

TOP RIGHT
Three crew members plastering a big bone in the Gobi Desert.

BOTTOM LEFT
Crew members finished the painstaking plastering process.

BOTTOM RIGHT
Jackets sitting around the truck that is going to take them to Ulaanbaatar.

from the specimen, the larger the tools that can be used; the closer you get to the bone, the smaller the tools that are used. A jackhammer is used to split the rock, and then the team switches to picks and chisels to remove layer after layer of sandstone. It's tough and dirty work, and the relentless sun and constant wind make it that much more difficult.

As the team switches from picks to rock hammers and brushes, it becomes apparent that the find is a half-metre-long foot bone (metatarsal) of a *Tarbosaurus*. In humans, the same bone is in the flat part of the foot (sole), and is usually no more than 5 centimetres long. However, to fully understand this specimen it will have to be unearthed and transported from this barren outcrop to a lab more than 1600 kilometres (1000 miles) away for preparation and study.

The most difficult part of the dig is removing the bone from the rock. It is critical to take away enough rock so that there is not too much excess weight for transporting, but not to get close enough to damage the surfaces of the bones. After hours of chipping at the rock around the find and brushing the bone clean, the fossil is put into a protective jacket. Once this is dry, the jacketed specimen is broken off the rock below the bone and is then flipped over. At that stage, the bottom of the specimen is also coated with a jacket of plaster and burlap.

Flipping the jacket is *the* crucial part. This is the time when the bone is being moved for the first time

in millions of years. The fragility of the bone is about to be truly tested, and if something goes wrong, the bone could easily shatter into thousands of minuscule shards and be lost to science for ever.

With the plaster dry on the top of the newly found *Tarbosaurus* leg bone, Currie and his team gather around for the flipping. It's a tense moment and should it go wrong, the afternoon's work in the scorching heat will have been for naught.

'Flipping a jacket is always stressful,' Currie says. 'Sometimes you lose everything you came for.'

Slowly, gently, they manoeuvre the specimen to see if it is loose enough. Then on the count of three they flip it over. It's a success. Small chips of bone have been lost, but the bulk of the bone is secure.

They cover the bottom half of the *Tarbosaurus* leg-bone with burlap and plaster, and, once it dries, it is ready to be loaded into the truck. It's another tense moment since the full package could go tumbling down the cliff side if they lose their grip. This one weighs in at 55 kilos (120 pounds) and takes three people to lift it.

Again, it's one, two, three, lift. And again, it's a success. The truck will now drive the jacket back to camp.

When funding allows, Currie has used helicopters to remove the heavy jackets, but this can be dicey as well. One year in Alberta, Currie and his team had dug up the hip of a duckbill dinosaur, preserved it in a plaster jacket and loaded it into a net below a helicopter for removal. The pilot was inexperienced and got caught in a crosswind. The fossil load began swinging like a pendulum, and the pilot couldn't pull out of it so he released the load. From a distance, Currie and his team watched the jacket fall to the ground. When it hit, they could see a massive dust cloud rise up. A few seconds later, they heard a crashing sound. They rushed to the scene.

'The plaster jacket had hit the ground so hard that it had rehydrated and turned to wet mush. And there wasn't anything inside, no bone or rock that was more than a few millimetres wide,' Currie recalls. 'It was completely destroyed.'

Digging up the *Tarbosaurus* bone is deemed a success, despite the fact that it is not part of a bone bed. But finding any ancient bone still gives Currie a rush.

'I've been hunting for dinosaurs since 1976, and the thing that amazes me is that as many dinosaur skeletons that I've seen, I can still get excited about them. I think it's so cool when you find a specimen and you realize that the animal lived at least 70 million years ago and you're possibly the first person seeing it. That's very exciting. For me now, it's the excitement of what information I can get from these specimens that will tell me about what these animals were like when they were alive. The excitement of discovering never really diminishes; it just changes in its nature a little bit because you're looking for slightly different things.'

* * *

Currie also visits the sites where tarbosaurs have been found in the past by other expeditions. There have been numerous finds and, though they are all single skeletons, there may be other skeletons nearby that have yet to be found. With the possibility that there might be more to be learned from these sites, each becomes another data point on the maps.

The first site Currie visits was discovered by a Polish expedition nearly 40 years ago. The original reports were rather confusing. In some places they said that the remains were of a *Tarbosaurus*, whereas in others they said that it was a *Saurolophus*, a duckbilled herbivore. To the untrained eye, the scraps of bone that remain in the quarry just look like a pile of bones, but Currie does not have an untrained eye.

He rummages through the bones. Combining the precision of a watchmaker and the determination of a forensic detective, Currie matches piece to piece. 'It's kind of like a puzzle, except half the pieces are missing, and the other half have been mixed up with ones that don't match,' he explains.

Suddenly, he comes to a realization. These aren't *Tarbosaurus* or *Saurolophus* bones – they are both. The two animals are in the same quarry. The bones also have about the same preservation, which means the animals probably died at about the same time.

Currie calls it an unusual occurrence because both the predator and the prey in the same quarry. However, because these are river environments, it's possible that the two bodies may have been washed together and buried at about the same time.

'When we find two specimens together like this, it's always possible that there was some kind of interaction between the animals that led to the death of both animals. There is one very famous example from Mongolia called "The Fighting Dinosaurs", where there's absolutely no question,' he explains. 'The claws of the *Velociraptor* are

'It's kind of like a puzzle, except half the pieces are missing and the other half have been mixed up.'

actually embedded in the side of the *Protoceratops*, and the *Protoceratops* is actually biting the arm of the *Velociraptor*. This just happened as they were being buried by sand in a sandstorm. But in other cases, when we find two skeletons of predator and prey associated like this, it's very difficult to say for sure what happened.'

Currie pulls out a red pen and writes directly on the bones where the specimens were found, what they are and what they represent, so there can be no confusion when they are studied back in the lab. If the specimen ever needs to be displayed, the pen markings can easily be removed using acetone or alcohol.

Almost certainly these two animals died at around the same time. Could it have been a predator–prey relationship? The *Saurolophus* could grow to 13 metres (42 feet) long. If this was the site of a kill, could the *Tarbosaurus* have done it alone? The answer is almost certainly not. Is there another *Tarbosaurus* skeleton nearby?

As Currie starts searching the area for evidence, he finds some very disturbing 'artefacts': plastic bottles, cigarette butts and an empty vodka bottle. These are the tell-tale sign of poachers. The plastic bottles are for superglue to hold the bones together, the cigarettes and vodka are the poachers' way of coping with the landscape. This site could have held another *Tarbosaurus* but it's impossible to tell for certain since poachers have destroyed it. The fossil sites of the Nemegt Basin have received government protection, but the funding isn't there to actually protect the site. In fact, the ranger tasked with overseeing security is located some 80 kilometres (50 miles) away – and he doesn't have a vehicle.

There's a constant threat of poaching in the Gobi, and each year more and more specimens are removed illegally and sold on the black market. Generally, the poachers go after the 'big ticket' items, such as claws, teeth and skulls. But in their desperate attempts to find these key bones they hack randomly at the specimens, destroying valuable scientific evidence along the way.

'As a quick and dirty way of getting these things out of the field so they don't get caught, basically what they do is just pour cyanoacrylate [a fast-acting adhesive] onto the bone so they can just rip it out of the hillside,' Currie says. 'The specimens do occasionally get seized by customs in Mongolia, and they're almost impossible to prepare afterwards because they're just covered in superglue. Most of the time the people who buy specimens on the black market are not as knowledgeable of what should be a good specimen or a bad specimen.'

Currie continues to examine the bone fragments. He is able to determine the size of the animals from the bones. There is an ankle bone from a *Saurolophus*, and the nasal bridge and toe of a *Tarbosaurus*. The ankle bone indicates that the *Saurolophus* was quite large, while Currie says the nasal bone is from a half-grown *Tarbosaurus*.

'The *Tarbosaurus* was probably less than half the size of the *Saurolophus* found in the same quarry,' he explains. 'The *Saurolophus* was probably in the range of 12 to 13 metres [39–42 feet], based on the few bones that we can see here. The *Tarbosaurus* was probably in the range of about 7 metres [23 feet].'

The bones will be packed and sent to the lab for further examination. If this was the site of the kill, the *Tarbosaurus* would have needed help. But there are no other skeletons around. Either they have been poached, or they didn't exist. But their existence together gets Currie thinking about behaviour.

'One of the most interesting aspects of studying dinosaurs these days is that we're learning something about dinosaurian behaviour and 10 or 20 years ago you never would have thought it possible to even conceive of what dinosaurs did. It was great to put it in a movie but, scientifically, you just had nothing to go on and you had no tools that would give you some indication of what the behaviour was like. But over the years, what we've found is that there are a lot of different kinds of fossil resources that can be exploited. You don't need to just go out and collect a single skeleton. What you can do is look at bone beds where you have many individuals that have died either over a long period of time, or just got washed together or died almost instantaneously and ended up in the same place at the same time. With tools that we have now and the knowledge that we have now, we can actually pick apart information from bone beds [and] show that in some cases the animals were associating in social groups.'

On the final night of the six-week Gobi expedition, the team are in an all-round celebratory mood. They cheer their success with a large dinner in their mess tent. The Mongolian vodka flows freely and each member of the team receives a toast. It's been a highly successful trip and the riches of the Gobi Desert have proved themselves once again.

The dino hunters recount their finds. Stacks of plaster jackets lie in piles next to the huge tent. No less than five different types of dinosaurs have been found. They've even found a whole series of footprints, including one belonging to the

Tarbosaurus, and they've also discovered what one of the dinosaurs ate.

Yuong-Nam Lee and his group found a *Therizinosaurus* skeleton with stomach contents. In its day, this distant cousin to *Tarbosaurus* must have been one bizarre creature. It had a small head, extra long neck, short tail and a large body, but the strangest part was its tiny arms with claws that were a metre long. This is the third therizinosaur partial skeleton that the group has discovered in the Gobi. The finds are important because they go beyond the claws that had been found previously, and therefore the scientists can now better determine what the entire animal looked like.

This is significant because the *Therizinosaurus* has been one of the most misidentified dinosaurs. The first *Therizinosaurus* find was first reported by the Russians back in the 1950s. It consisted of a couple of huge claws more than a metre long, and they incorrectly identified it as a turtle. The thing was so bizarre-looking that it took a long time to work out that it was a dinosaur. 'Once scientists figured out it was a dinosaur, nobody was sure what kind of dinosaur it was,' Currie says. 'The claws looked kind of like theropod claws, but the bones in other parts of the skeleton did not seem to have come from theropods.'

Eventually it was determined that there was another weird type of dinosaur found by the Andrews expeditions back in the 1920s, which had also been misidentified. Originally, the dinosaur was identified as a tyrannosaur because a tyrannosaur skeleton had been mixed up with some therizinosaur arm bones. The error was discovered in the 1980s, and the bones were separated into their respective animal types. Around the same time, one of the Mongolian palaeontologists described some specimens from that country that he identified as a new family, failing to make a connection with *Therizinosaurus*. In China, another specimen was found and identified as a sauropod. Next, an American palaeontologist named Greg Paul looked at them, and he thought that they could be prosauropods, which are a very ancient group of dinosaurs that disappeared before the Late Cretaceous. He also identified it as a plant-eating dinosaur. The breakthrough came in the late 1980s when some specimens in China were found, and Dale Russell finally put it all together. He determined the creature was a theropod and at some point it was associated with these so-called turtle claws from Mongolia.

'Even though they are related to *Tarbosaurus*, a lot of people have postulated that these are actually plant-eaters. They make a comparison to giant ground sloth which have claws similar to this but use them for pulling down trees so they can eat the leaves,' Currie says. 'In all probability, these weren't doing that. Because they were related to *Tarbosaurus*, they probably were still carnivores, but were very specialized and maybe those claws were used for fishing.'

The most incredible aspect of Yuong-Nam's therizinosaur find is that the scientists have found out what it used to eat. Many dinosaurs had teeth that were not very good for chewing so, like modern-day birds, they would swallow rocks to help them break down the food in their stomachs. They also had long, slender bones called gastralia to support their stomach walls. Yuong-Nam found the stomach bones, and inside the walls were small, very smooth rocks that are clearly not from this part of the desert. Within this stomach, there were even the fossilized skeletons of fish. 'From these finds, we now know that the animal was eating fish, and we know what the rest of the skeleton looked like, so we will publish a paper on that,' Currie says.

Yoshi Kobayashi was working in a quarry with footprints of ornithomimids or ostrich-like dinosaurs. He is hoping the discovery will fill in the blanks about how that dinosaur lived and its biology.

Dave Eberth had a breakthrough in dating the land. 'We've got almost 60 years of conflict and misunderstanding of how these fossil localities across Bugin Tsav and across the Nemegt Basin all relate to each other,' he says. 'This summer, one of the things I was able to do was to identify a couple of marker beds that we found in localities that were more than 100 kilometres [62 miles] away, and all of the localities across that transect over a distance of 100 kilometres, we were now able to tie together. This is extremely exciting because it is going to enable us to test our ideas about which dinosaurs were living there before which other dinosaurs, and which ones went extinct before which other dinosaurs.'

Louis Jacobs and his group found only the second ever skeleton of a *Barsboldia sicinskii*, a duckbilled dinosaur. It was named in 1981 in honour of the granddaddy of Mongolian dino hunters, Rinchen Barsbold. The specimen found by Jacobs' group is complete from the tail to the hips, and more of the skeleton continues into the hillside and will be excavated in the future. Though it was enormous, *Barsboldia* was probably a prey animal for the even larger *Tarbosaurus*.

'That dinosaur is big enough to stand flat foot and shit in a dumpster!' Jacobs says enthusiastically. 'The larger the animal, the stronger the animal. It

'It's not unreasonable to think that there was a social interaction that led to tarbosaurs being well fed.'

also follows that the larger the predator, the larger the prey in general the predator can take. You don't see lions spending their time catching mice, so you can expect that *Tarbosaurus* would go for relatively large prey, and *Barsboldia* and duckbill dinosaurs are large prey. But full adults are strong large prey. I would expect that the younger or the weaker or the ones that were less able to be aware would be the ones who would fall prey to *Tarbosaurus*.'

The idea that tarbosaurs might have hunted such large prey as *Barsboldia* opens up questions of cooperative hunting behaviour. 'If you have a large number of top predators, they have to be sustaining that population,' Jacobs says. 'So either there's a lot of prey and nobody is around each other and they're all going their own way, or they have a social structure that allows some sort of, if not interactive, cooperative behaviour, that manifests itself in getting the food. It's not unrealistic to think that you have a group of tarbosaurs – a gang, if you want to call them a gang – that goes after prey when the prey is there. Let's say they're eating *Barsboldia*. If you've got a concentration of *Barsboldia*, where are the predators going to concentrate? Right there with them. You could expect there to be a social interaction. Whether it has the sophistication of the gang in *West Side Story*, I don't know, but it's not unreasonable to think that there was a social interaction that led to them being well fed.'

Currie was primarily focused on the parts of four tarbosaurs that the team found and what they might tell him back in the lab. There are also three more prospective sites that need to be examined the following year. But Currie is most excited about his re-examination of two of the *Tarbosaurus* specimens that were found within a couple of metres of each other: one is an adult specimen and one is a juvenile. They could be mother and baby.

'I don't think there's any question that at Bugin Tsav we're looking at maybe a dozen tarbosaurs that were buried within days or possibly even hours of each other,' he says. 'Whether it represents one group of a dozen *Tarbosaurus* or several smaller groups of tarbosaurs in a way it doesn't matter, because you've got these animals moving in packs of some size.'

Currie is convinced that the concentration of *Tarbosaurus* in certain places at certain times and the fact that the prey animals available to them were so large have advanced his dino gang theory. 'This was the evidence I needed to take an idea much further ahead than anybody thought it possible before,' he says.

As notable as these finds are, there is always more on the wish list. Currie says his ultimate find would be a mother *Tarbosaurus* with a baby, and that both of them were feeding on a duckbill dinosaur at the time that they died. This would show cooperative hunting at its most sophisticated.

'These things all sound very improbable, but the reality is that under very, very special circumstances sometimes you can find these things preserved,' he says.

As the sun comes up the next day, the jackets are loaded into the back of huge trucks. There is one last stop they have to make to be securely packed up before being shipped to the lab. The jackets are driven to the outer suburbs of Ulaanbaatar, the capital of Mongolia, a strange combination of post-Soviet buildings, many with gers, the traditional nomad tents, in the backyard. This is where the compound is located, behind a set of huge steel doors. Inside are the massive ex-army, Russian trucks that are used in the Gobi expeditions. The jacketed fossils are placed in locked shipping containers. This is nerve-wracking as the specimens are dragged between trucks and re-stacked and made ready for the last leg of their journey. Dust flies and sunlight streams into the small confined space as the team heave the jackets into place.

The fossils now have another lengthy trek in front of them as they travel by land and sea to the brand new high-tech research facility at the Hwaseong Dinosaur Laboratory in Korea. 'Very often people think that the excavation is the end of the deal, but it's not,' Currie says. 'It's just the beginning. In fact, it's the shortest part of the whole operation. Of course finding the specimens can take a long time, but once you've found them, even to excavate a major skeleton will take you a maximum of a few weeks. In terms of the preparation and the research, you're talking years.'

chapter 6

to the lab ... and beyond

If collecting dinosaur bones is a long and slow process, examining them is an even longer and slower one. Specimens shipped from remote locations take months to reach the lab and years to fully study. It is in the lab that what has been unearthed is thoroughly examined using the collective wisdom of the scientists and the most advanced scientific techniques, and then compared to the historical record. An experienced palaeontologist comfortably makes the transition from the rigours of fieldwork to the intricacies of the technical study employed in the lab.

'Pulling the fossils out of the ground is critical, but it turns out that laboratory studies help us to flesh out what the animals are really like,' anatomist Larry Witmer explains. 'Only when we get some of the data back into the laboratory can we really start to sort out what's going on. Sometimes when you take fossils out of the ground, we just get a glimpse of what they might be telling us. It's really only when we get them back into the lab and we can really study them in detail that we learn what the fossils are really telling us about the living, breathing animals.'

As much as Phil Currie loves fieldwork, his first real experiences in palaeontology came when he volunteered in the lab at the Royal Ontario Museum while in college. He actually chose to attend the University of Toronto so he could be involved in an active lab. He currently has his own lab just across the hall from his office in the Biological Sciences Building at the University of Alberta, where he and his students work on dinosaur material collected in Alberta.

To examine the finds from the five-year Gobi Desert expedition, Currie must travel to the Hwaseong Dinosaur Lab in Hwaseong City, Korea, which is run by the expedition's leader, Yuong-Nam Lee. This is the next destination for the specimens that were collected in the desert and the place where they will be re-assembled and examined over time. Ultimately, the specimens will all return to Mongolia. In the autumn of 2009, the scientists who were in the field in the Gobi Desert – Currie, Yuong Nam-Lee, Yoshi Kobayashi, Eva Koppelhus and Louis Jacobs – gathered for the first time together in a lab. It is at this brand-new facility that the scientists aim to reveal new details about the dinosaurs' anatomy and, most importantly, their behaviour. The bones will be divided and catalogued and then the group will determine the most appropriate person to take the lead in studying them, and the pieces of a mystery will begin to come together.

'This is a very exciting part of the process,' Currie says. 'It's planning for the beginning of the results in a sense. The results for us are the scientific publications and we don't see those for many years. But starting to put all those pieces together in the jigsaw puzzle and to solve some of the problems that you've developed while you were collecting the materials, that's a really exciting part of the process.'

Louis Jacobs explains that the lab is a place to step back and study the materials in a controlled environment. 'When you're out in the field, you're

Yuong Nam-Lee's dinosaur preparation laboratory in Hwaseong.

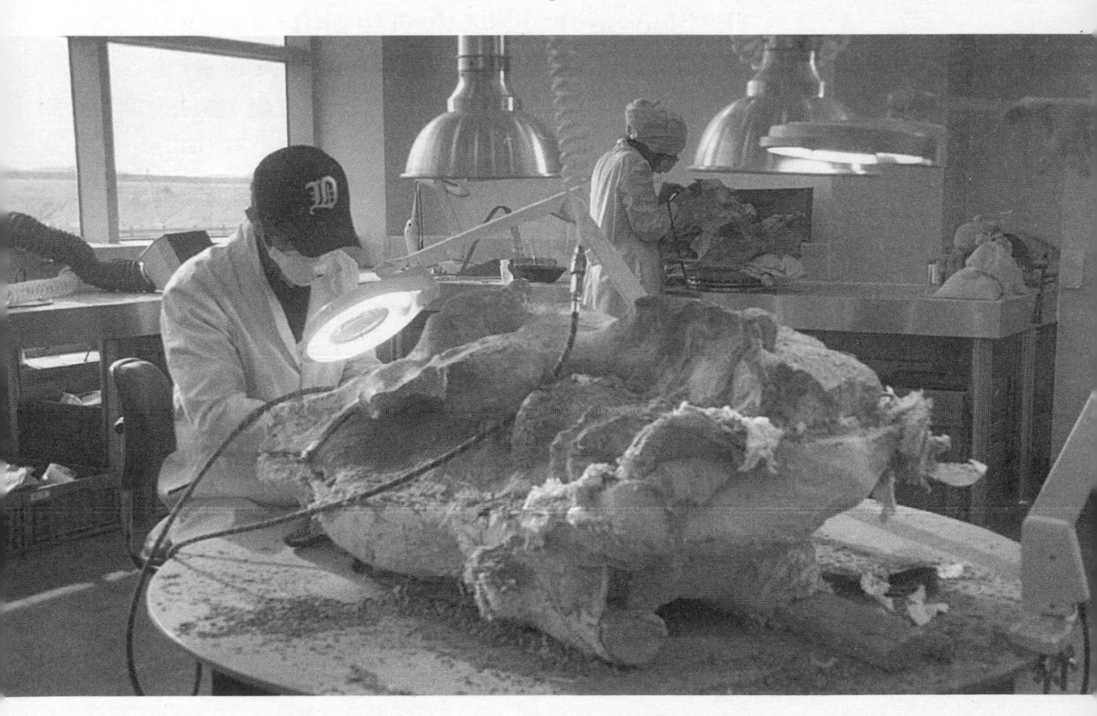

putting up with the heat and you're putting up with the dust, you're putting up with long hours, and that's a lot of physical effort involved. Plus, you can't see everything about the bones when you're digging them up and covering them up with plaster. You don't even want to expose them all the way because you don't want to damage them. It's very understandable that your first impression of what you've got may not turn out to be what it is. So you come back here and you open up the packages and you look at them, and sure there's lots of surprises, good surprises, because they turn out to be things that you just didn't even know you had or the quality of what you have.'

The dinosaur bones collected at Bugin Tsav during the past four field seasons are housed in a storeroom. When the scientists arrive to begin examining them, they are still wrapped in their protective jackets. There are hundreds of beautiful surprises to be opened. Though the amount of work is vast, when the group gathers the energy level is high.

'It's kind of like Christmas,' Currie explains. 'You finally see the specimens in the lab. It was months ago that we shipped off these specimens. Many of them were wrapped up in paper, plastic or even plaster. Then when you finally see them, you can actually take them out one at a time, lay them out on a table, and see what you managed to accomplish. First of all, some of the specimens you don't even remember. And secondly, you're very surprised by the quality of what you've collected. It's a bit of a surprise when you see the material laid out, because when you're collecting it, you're collecting one specimen at a time and, as quickly as possible, you're trying to wrap them up and get them boxed so that they're safe. When you finally unwrap them all and lay them out, you're very surprised because you never saw it all together before.'

Preparing specimens for examination is an exhaustive process. The plaster jackets that protected the fossils during transport are carefully opened with a small circular saw, in much the same way that a plaster cast is removed from a broken arm. Once all the plaster is removed, the remaining rock must be chipped away from the bone. Only when all of the rock is removed can the bones be laid out and examined in detail. To the trained eye, it's a window into a world that vanished at least 65 million years ago.

'When you look at these bones, sometimes you forget these represent an animal that was at the height of its evolution,' Currie says.

One of the first jackets to be opened is a *Tarbosaurus* hip bone that was discovered on the Gobi expedition in 2008. The excitement is palpable, but once the top of the jacket is off, it is clear that there is still a lot of work to do. The team must now separate all of the rock from the broken bones, which must be consolidated and glued back together. Once this is complete, they will see what the find reveals.

Bones help reveal the body mechanics of the *Tarbosaurus*, and the scientists compare the ratios of the leg bones and the action of the movement. By looking at the hips and vertebrae, the scientists can calculate how big and how heavy the dinosaur was. Tyrannosaurs walked on their toes with the sole of the foot raised off the ground the way most animals do today. (Humans and bears are flat-footed exceptions.) But the size and weight raise questions: if *Tarbosaurus* weighed up to 6 tonnes, how fast could it have really run? Would all that tonnage drag it down, causing it to lead the life of a sedentary lizard? Or did it have such powerful muscles that it could run at great speeds? The answers lie in how *Tarbosaurus* grew.

Though a comprehensive analysis on the life-cycle of a *Tarbosaurus* has yet to be done, Currie believes they had a similar life span to an *Albertosaurus* or a *Tyrannosaurus*. The oldest *Tyrannosaurus* specimen is estimated to have been 28 years old when it died. 'The timeline of their growth is very much like humans,' Currie says. 'These things come out of the egg and are probably growing like crazy. We know this from *Albertosaurus*, because we have a two-year-old *Albertosaurus*, and a pretty small, two- to three-year-old *Tarbosaurus* that was recently described. But then when they hit their teenage years, the growth rate increases even more and goes really crazy. They hit their highest growth stage, depending on the individual, between 12 and 14 on the lower end. By the time they are 19, the growth rate levels off and basically they add more mass and get fatter. So at around 20 years old, they start bulking out and that's going to slow them down. The proportions and your mechanical ability to run fast [are] controlled entirely by how big you are, how heavy and how long you are.'

The bones from the *Albertosaurus* bone bed that Currie discovered in Canada show a full range of growth for one member of the tyrannosaur family. Normally, when a dinosaur species is found, there are a large number of adults. Because the adults are so massive, when they die they don't get eaten or destroyed all that fast. The bones are heavy so they stay together. Therefore, palaeontologists have a much better idea of what's going on with adults than they do with juveniles. There are no really good baby tyrannosaur specimens of any genus or species – no *Tarbosaurus* babies or *Albertosaurus* babies. However, there are partial skeletons. The *Albertosaurus* bone bed in particular has a lot of partial skeletons including the feet of these animals.

'What we see in one single bone bed is that the proportions change remarkably from the juveniles to the adults,' Currie says. 'This shouldn't really be a big surprise because we know the same thing happens with humans or with dogs. We know for example that a puppy with big feet is going to become a very large dog. Well, we can say the same thing for some of these dinosaurs.'

In his career, Currie has seen *Tarbosaurus* skeletons of varying sizes and ages. He is struck by the fact that as they grow from hatchling to teenager to adult, tarbosaurs don't just change size but also body shapes. The relative lengths of the metatarsals, tibia and fibula change as the animals grow. When the length of the thigh bone is compared to that of the foot bones, the foot bones are two-thirds the length of the femur in the juveniles. The bigger the animal gets, the smaller

'As they grow from hatchling to teenager to adult, tarbosaurs don't just change size but also body shapes.'

these foot bones are in terms of relative length compared to the other leg bones.

'When we look at the whole range of size in *Albertosaurus* bones, we see two things happening,' Currie explains. 'Number one, the juveniles' metatarsal bones are very long. They're almost as long as those of more mature animals, but they're very slender. The second thing is that when we compare the length of [the foot] bones to the bones in other parts of the legs, say the upper-limb bone [the thigh or femur], we see that in the juveniles the bones from the flat of the foot, the metatarsals, are almost as long as the upper-leg bone. And yet when they get to be a large adult, the proportions change dramatically.'

The flat of the foot bones of an adult *Tyrannosaurus* are not very much longer than those of an *Albertosaurus*, but they're much thicker and much more massive than the *Albertosaurus*, and they're almost half the length of the upper-leg bone.

'It's telling us that in terms of leg proportions, these animals are much more lightly built as juveniles, and they become much heavier, more massive and slower as adults. That's more or less what we would expect to see because it's what we see in modern animals too.'

A tarbosaur leg is huge, measuring 2.5 metres (8 feet) long, and it would have been extremely muscular. But stripped to the bare building blocks, the proportions of the teenage *Tarbosaurus* leg were very close to those of the fastest two-legged modern runner, the ostrich. The *Tarbosaurus* leg bones show a similar size ratio to the ostrich leg, and the muscle attachments appear to be almost identical, thanks to innovative computer-imaging techniques that allow the scientists to examine the muscle structure of both the modern birds and the ancient dinosaurs legs.

'When you look at modern birds, particularly big birds like ostriches, and you do a comparison with some of these large dinosaurs, it really is shocking,' Currie says, 'because, fundamentally, you are looking at almost exactly the same thing. These animals are relatively slender in the lower part of the leg, compared to the upper part of the leg which is much more muscular. The proportions are quite similar with ostriches so we can make direct comparisons.'

Ostriches have been compared to a certain group of dinosaurs called the ostrich-mimic dinosaurs – ornithomimids – which had very long legs and similar proportions to ostriches, causing scientists to believe they must have been able to run very, very fast. In fact, the ostrich-mimic dinosaurs have exactly the same proportions as young *Albertosaurus* that are about the same body size as them. 'That strongly suggests to us that young *Albertosaurus* was an animal that was much faster and more manoeuvrable than the adult forms,' Currie says. 'In fact, it may have been the fastest, most manoeuvrable dinosaur for a certain period of its life cycle, and that strongly suggests that it's doing something a little bit different than the adults.'

The comparison between ostriches and dinosaurs is especially apt because birds are the direct descendants of dinosaurs, so there is an existing relationship between them.

Because we know that ostriches are fast, this comparison leads to a debate about how fast large dinosaurs ran. The examination of the dinosaur bones in the lab and their similarities to ostrich bones will lead Currie back to the field. Only this time, he isn't looking for fossils. To further his study, Currie is going to use progressive techniques to study the ancient beasts.

* * *

Currie knows what the *Tarbosaurus* looked like and how its bones were structured, but in order to see how it moved, how it killed and how it behaved, he feels that he must compare this amazing ancient predator with modern-day counterparts living in the wild today in a bid to understand its social behaviour, its character and anatomy, and ultimately its hunting strategy. This is a compelling way to study the ancient beasts, and one that Currie believes is critical to paint a complete picture.

'The idea of comparing modern animals with older animals is not a new idea,' he says. 'It has been around since the late nineteenth century. Yet to a large extent it's really been since the 1980s and 1990s that people have thought more of doing [it]. In the very general sense, people would look at ancient sediments and talk in terms of what modern animals or modern environments would do, but nobody actually followed through by making the comparisons. Once people started looking more at ecology in the 1960s and 1970s, then people started thinking in terms of these ancient ecosystems in terms of modern ecosystems and started making those kinds of comparisons.'

Initially, scientists relied mainly on the fossilized bones to study dinosaurs, but Currie points out that even with the most complete bone beds, it takes years to sift through the evidence.

'Behaviour is that much more difficult to interpret from fossils and consequently you have to have so much evidence and so many lines of evidence. Working on bone beds has been a great boon to this whole development, but bone beds take years to excavate, and you are not actually going to see the results for maybe 10 or 15 years. It's a slow process, and given the fact that there were very few palaeontologists working on dinosaurs until quite recently, it meant that there weren't a lot of ideas out there. People wanted to make the comparisons, but it wasn't happening.'

Where the new techniques for studying dinosaurs have opened once-closed doors to palaeontologists is in the area of behaviour. In the early days when scientists compared ancient animals with modern

'The study of dinosaurs requires examining not only the ancient past but also the present.'

animals, there was very little follow-through and therefore progress in our understanding of areas such as behaviour was far slower than it is today.

'We used to think that some things were out of reach,' Larry Witmer says. 'Behaviour was just too much speculation. However, we have some new ways of looking at it. We never used to think that we could ever come up with the behaviour of an animal from 70 million years ago. We are opening up new vistas for reconstructing the behaviour of extinct animals. We can now actually reconstruct what the bones are telling us about behaviour. We can bring them back to life. What's remarkable is how rapidly our views are changing about tyrannosaurs. We used to think of tyrannosaurs as upright, slow moving, dull-witted animals, but science studies show they are much more horizontally oriented, much more fast moving and relatively clever animals.'

Currie is among the palaeontologists who believe the study of dinosaurs requires examining not only the ancient past but also the present. 'There is no question that the present is the key to the past,' he says. 'If you're a palaeontologist and you're working on fantastic animals that don't exist any more, basically if you don't understand what is going on in the modern world, you're not really going to understand what was going on in the ancient world either. Sometimes we think that the ancient world is very different from today, and yet in many ways, the play has always been the same and what we're changing is the actors over time. I don't think dinosaurs did anything unique; they did it very well at that time but the same processes are going on, the same play is still being enacted today.'

In the autumn of 2009, Currie examined three present-day animals to help him explain the story of dinosaurs that lived more than 70 million years ago: the Komodo dragon, the ostrich and the lion. The results of these studies would both test the core of his dino gang theory and surprise Currie himself.

chapter 7

speedy creatures

ecause of the massive size of most dinosaurs, many people assume they were slow. But, given that the shape and musculature of the legs of a juvenile *Tarbosaurus* are similar to those of an ostrich, could the *Tarbosaurus* have been a speedy predator? To bring the calculation of a dinosaur's speed to life, Currie will rely on both the most up-to-date techniques and an internal and external examination of an ostrich.

Dinosaur speed analysis is primarily done in two ways. The first way is by using the length of the legs, the proportions of the various leg bones, the length of the body, and the estimated body weight of the dinosaurs, and comparing these metrics to similar-sized living animals.

The second method is by using trackways of fossilized dinosaur footprints, which are known from as far back as the appearance of the first dinosaurs 225 million years ago. Employed since the 1980s, this methodology involves measuring the footprint length and the distance between tracks of living animals, and comparing them with trackways of extinct animals. The greater the distance between the footprints, the faster the animal was running. After the distance of a stride (the distance between two consecutive footprints from the same side of the body) and the average lengths of the footprints are measured, a mathematical formula is used to compare these metrics with the observed speed of the animal. Substituting the first two measurements from dinosaur trackways into the formula can then give an estimate of the speed that the dinosaurs were walking or running.

'The holy grail would be to find a dinosaur dead in his trackways,' Currie says. That, of course, has not happened.

Currie first wants to examine the ostrich close up because of the similarities of the bone lengths and structure between the flightless bird and extinct dinosaurs. 'A baby tarbosaur has about the same build as an ostrich,' he says. 'And so if you want to understand what a baby *Tarbosaurus* was capable of right through into its teenage years, it's probably good if you understand what fast-moving animals like ostriches are capable of.'

In order to determine whether a young *Tarbosaurus* was as fast as an ostrich, Currie travelled to the Tswalu Kalahari private game reserve in South Africa to observe these birds in their natural habitat. Located at the edge of the Kalahari Desert, the Tswalu Kalahari is South

Today's ostrich appears to run much like a Tarbosarsus *did.*

Africa's largest private game reserve covering some 670,000 hectares (2,600 square miles). The relatively intimate nature of the reserve allows Currie to observe the ostriches in their natural habitat with minimal distractions.

There was nothing high tech about measuring the ostrich's speed here. Currie's guide drove him out into the savannah in a jeep and raced alongside the running ostriches. It was a hairy ride along a dirt road with dust and rock spraying everywhere. The speedometer read 40 mph. In addition to gauging the speed, Currie also wanted to witness how an ostrich actually runs. 'It's pretty amazing watching ostriches run because they are so much like tarbosaurs in terms of size and maneuverability,' he says. 'The amazing thing is that they are so manoeuvrable.'

When scientists study modern animals to find out how fast ancient animals could move, they really only have to look at their hind legs. This holds true for ostriches, as well as speedy gazelles, because the similarities in their leg proportions tell scientists that these are fast animals.

'As a comparison we can look at ourselves and ostriches,' Currie says, bringing the point closer to home. 'An ostrich is an animal with very long legs, much longer than our own legs compared to its body size, so the long legs themselves tell you that this is probably a very fast animal.'

There are also other major differences between ostriches and humans. We are flat-footed and walk with what's called the metatarsal bones (the bones in the sole of our feet) on the ground. Additionally, the proportions of the leg bones are different.

'If you look at an ostrich, though the ostrich is in fact running on its toes and the bones from the sole of the foot, the metatarsals, are elevated into the air, they've become an extra joint and that joint is incredibly long,' Currie continues. 'It is almost as long as the lower part of our legs and that gives the ostrich a bio-mechanical advantage in running that we just don't have. There is no way a human being, even on steroids, could ever out-run an ostrich.'

Ostriches can reach speeds up to 45 mph – twice as fast as an Olympic sprinter. Is it possible that *Tarbosaurus* could have reached those speeds? If so, it would be a mesmerizing prospect: 6 tonnes of muscle capable of chasing down virtually anything it wanted.

'People don't think of tyrannosaurs as being fast animals,' Currie says. 'They're just such big, massive animals in terms of what they're familiar with; but they forget that every tyrannosaur was a baby at one stage and it was immature for a long period of time. Basically, a tyrannosaur had the same kind of life cycle that we do. They didn't grow up until they were in their late teens and twenties, and up to that point in time these animals were relatively slim, lightly built, and very, very fast and comparable to ostriches. When you're looking at an ostrich-sized tyrannosaur, you're looking at an animal that would have run faster than anything that was around at that time. This thing was hell on wheels.'

Based on their size and bone structure, it has been estimated by some scientists that tyrannosaurs, and therefore tarbosaurs, could reach speeds of anything from 25 mph to 45 mph. But to arrive at a more accurate estimate, Currie needs to know more about how their muscles worked. He needs to get inside the ostrich leg. For this, he travels to the nearby office of veterinarian Jan Vorster to dissect an ostrich.

'Trying to understand [dinosaurs] collectively and what they were capable of in terms of running and manoeuvring is very difficult because they are dead animals,' Currie says. 'If we understand what [the] musculature is doing on modern animals, then we can interpret which muscles attach to each one of the landmarks on a tyrannosaur bone.'

* * *

Unlike bones, muscles are not preserved as fossils, so an ostrich leg muscle is the closest thing to a *Tarbosaurus* leg muscle that Currie will ever see. He has done an ostrich dissection before, though not with a fresh specimen.

'Most of the time when we do the dissections, we have animals that are in formaldehyde, embalming fluid or preserved with alcohol, and they just don't perform the same way,' Currie says. 'In the case of a fresh one, we could actually take the muscles, isolate the muscle masses, flex the muscles and make the muscles move. So you could see that by pulling a certain muscle by the hip that you could make the foot move, and you could get a sense of not only how big the muscles were but also the tendons that were controlling them.'

During the dissection, Currie picks up the ostrich leg and flexes it to check the spring tension for himself. He then watches as Jan Vorster cuts the tendon away from the bone, allowing him to see the attachment point. Dissecting the hind limb gives Currie a better sense of mass proportions and how the hind limb on a tarbosaur might have worked. It is clear that the ostrich, like the tarbosaur, did not have too much weight in the lower part of the leg. That part of the body would not have been so easy to move if it was heavy, so the muscles that control that part of the leg are concentrated in the upper part of the leg and powerful tendons connect the two regions.

'Of course, a dinosaur isn't exactly like an ostrich, because there's been a lot of modification in any bird leg, including that of an ostrich,' Currie says. 'But the upshot of it is that they do come from the same lineage. Understanding how these tendons work, in particular their actions on the lower part of the foot, shows how the proportion changes in the leg are significant.'

From a tarbosaur bone, Currie can tell where many of the muscles were attached, though he can't really determine the shape of muscle. The ostrich had a 2.5-centimetre (1-inch) thick tendon, and Currie believes that the *Tarbosaurus* tendon would have been proportionally large. He is able to make a comparison between what he sees on the ostrich and what the tendon would have been like on the tarbosaur.

'It helps to understand muscles and the way they insert onto the bone,' he explains. 'In many cases, you have a big attachment point on a straight bone, and this forms a lever so the muscle can control things mechanically. But in other cases, a muscle just comes down and dissipates into a fascia that covers the bone so you don't actually see any attachment for the muscle.'

The muscle system in the back of a tarbosaur leg was as complex as that of the ostrich. Scientists can look at a tarbosaur leg bone and see the scars where the big muscles attached, as well as some of the levers on the bone. However, most of the attachment points, especially for the smaller muscles extending from the hips and tail to the lower part of the leg and foot, cannot be seen. By examining an ostrich leg Currie can gain a clearer understanding of what was likely to be happening with tarbosaurs.

'The amazing thing to me is that when you look at these joints they really aren't different at all from the dinosaurs,' he says. 'You could put them side by side, and maybe have a little trouble identifying the differences.'

In the old diagrams of a *Tyrannosaurus rex*, the legs appeared to be almost as massive as those on an elephant. 'But the reality is that like an ostrich, these animals had the bulk of the muscles concentrated up top,' he continues. 'And once you get into the lower part of the leg, everything is run by tendons and ligaments that give it a certain amount of spring.'

The key to speed lies in the tendons. They function like enormous rubber bands – the tighter you pull them, the more they spring back. It is this spring that maximizes the propelling force; thus greater tendon tension generates a more powerful stride. The leg proportions are crucial. The longer the metatarsals, the more spring there is in the tendon and the faster the animal can run.

Currie concludes that the similar proportions between the ostrich and the tarbosaur and the long, springy tendons mean that a young *Tarbosaurus* would have been capable of considerable speed.

'The leg proportions are so similar to what we see in ostriches that you can't escape the conclusion that a half-grown *Tarbosaurus* was an incredibly fast animal. Start picturing tarbosaurs as running together in flocks similar to groups of ostriches. Or baby tarbosaurs trying to run down prey, or scaring them towards the adult tarbosaurs that could do the killing. Then I think you start to get a picture of how incredibly fast these animals must have been.'

From the study of the bones and the dissection, Currie is able to conclude that young tarbosaurs would have been moving faster than a world-class sprinter. 'I think they were capable of hitting some pretty high speeds, maybe not as fast as an ostrich but faster than any of the other dinosaurs.'

But what about adult tarbosaurs? Unfortunately, there is no two-legged animal alive today that can be directly compared to a grown *Tarbosaurus*.

'There's no question that size has a cost,' Currie says. 'As tarbosaurs got bigger and bigger, in spite of the fact that they had these ingenious adaptations to reduce the amount of weight in the lower part of the leg, they still had to compromise. To carry the increased weight, they needed to beef everything up, and it was for this reason that an adult *Tarbosaurus* does look very different from a juvenile. But the juveniles could still run much like ostriches.'

The only problem with the dissection is that it had to be performed in a single morning before the specimen went off, and by the time Currie and Jan Vorster were finished the specimen was already starting to reek. 'It was pretty overwhelming,' Currie says, laughing. 'In addition to that, the thing was covered with live ticks. And I *hate* ticks.'

To determine how fast an adult *Tarbosaurus* could run, Currie sought help from the foremost lab of its kind, located in the Royal Veterinary College in Hatfield, England. There, scientist John Hutchinson has developed new ways of estimating *Tyrannosaurus* running speeds. Hutchinson uses a special motion capture process to create an accurate model of motion and speed. His high-speed cameras have filmed everything from horses and cheetahs to humans. From this he can provide estimates of speed for different-sized dinosaurs.

'All we have for a dinosaur is largely the bones and occasionally some footprints or other things,' Hutchinson says. 'So how can we reconstruct what they might have behaved like? The laws of physics were the same back then as they are today. Gravity was really no different, and animals are generally made of the same stuff – muscle is muscle, bone is bone. So we can apply what we know from living animals and the laws of physics to dinosaurs and reconstruct them within certain bounds of possibility.'

Hutchinson starts by making the obvious comparison between humans and *Tyrannosaurus* because they both ran on two legs.

'The great thing about humans is they're bipedal just like *Tyrannosaurus* was. Regardless of big, major differences in their anatomy, there still should be some fundamental similarities in how a human runs and how a tyrannosaur would run as well. So we can learn a bit about *Tyrannosaurus* from studying humans, birds and other bipedal animals.'

The process of measuring a human's speed involves placing reflective markers on to a runner and having him run past high-speed video cameras that film at about 250 pictures a second, which is about ten times as fast as your usual camera. At the same time, the runner goes over force platforms that register the pressure of his leg pushing against

'There still should be some fundamental similarities in how a human runs and how a tyrannosaur would run.'

the ground with each stride. The data is collected by a computer, thus allowing Hutchinson to observe what the runner is doing at any instant.

'Using our markers, we can measure exactly how every part of his body is moving, so different parts of the body move at different speeds and in different ways,' he explains. 'For example, the foot is moving quite quickly during running when it's off the ground, whereas upper parts of the leg are moving a little more slowly. This is true of all sorts of different species. So we can really measure his motion in great detail and understand how all the different parts contribute to the motion of the whole.'

There are certain principles of movement that apply across a wide variety of species. All two-legged runners in the animal kingdom exert two and a half times their body weight of pressure. That means a 100-kilo (220-pound) runner's leg must support 250 kilos (550 pounds) of force each time his foot lands. For an adult *Tyrannosaurus* that would be an incredible amount of force.

'A large *Tyrannosaurus* would be exerting a force of around 12, maybe even 15 tonnes per step on each limb,' Hutchinson says. 'To prevent the leg from actually collapsing under that force, the leg muscles have to work really hard. You need really big leg muscles to prevent the limb from collapsing.'

Currie points out that this fact had a huge effect on the shape of the adult's leg and, ultimately, its top speed. 'These increased disproportionately and the bigger you get, the thicker you become – which means you become even bigger yet,' he says.

From these enormous forces, Hutchinson has been able to calculate that *Tyrannosaurus* did not walk straight-legged or crouched low, but rather ran half-crouched to absorb these forces. For every 50 kilos (110 pounds) a growing tarbosaur added from teenage years to adulthood meant that an extra 125 kilos (275 pounds) was added to each running footfall. Putting all the data together, Hutchinson

At the Royal Veterinary College, scientists measure the pressure of a human running stride to determine the exertion of a Tyrannosaurus.

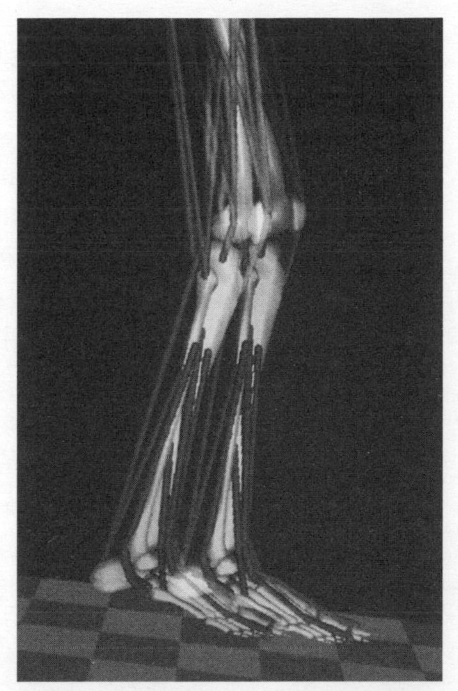

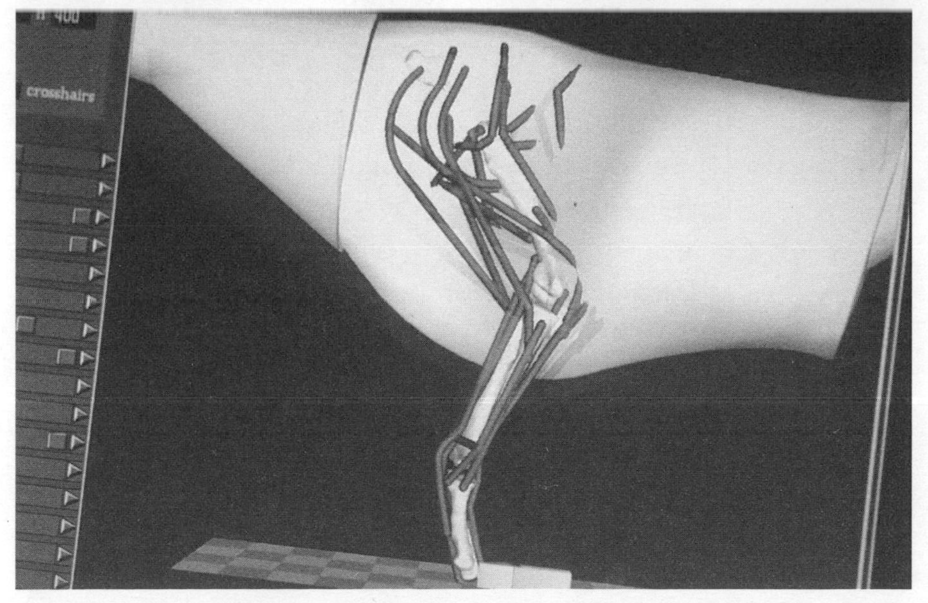
crosshairs

estimates, 'A large tyrannosaur might have been able to run somewhere between 15 to 25 miles per hour.'

Using the information from the ostrich study, that would indicate a decrease from an estimated 45 mph to a top speed of 25 mph. Hutchinson, however, is not certain that younger tyrannosaurs could achieve such speeds, though he does believe they was far more agile than the adults.

'I think there's something to the idea that younger tyrannosaurs could have been more athletic than larger tyrannosaurs,' he says. 'There was a huge thigh change during their growth from an animal that was only 45 kilos [100 pounds] to an

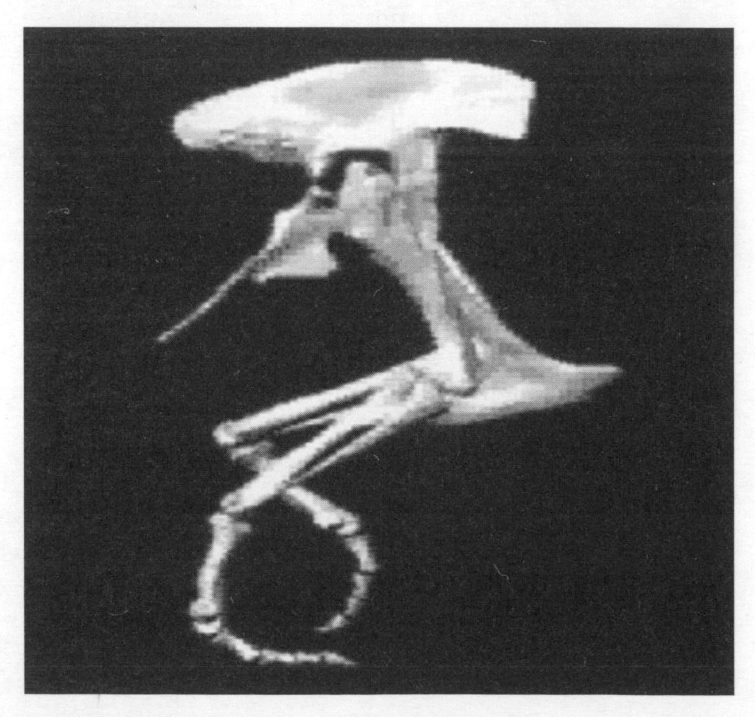

A depiction of a running Tyrannosaurus.

animal as heavy as an elephant or more. It's still not certain if a small tyrannosaur would run faster in absolute terms like miles per hour compared with a big tyrannosaur. But it is very certain that they would be more athletic, more bouncy, and able to do certain kinds of activities that a big tyrannosaur just couldn't.'

Because this is similar to modern mammals (the small are fast but get slower as they grow older), does this mean the adult tyrannosaurs were too slow to hunt in gangs, or to pursue and overtake their prey?

Not necessarily, but it does means they were not a pure pursuit predator. The slower speed implies the gang might be effective ambush predators – with the younger faster tyrannosaurs chasing and driving prey into the mighty jaws of the slower yet much more powerful adults.

'I certainly agree that smaller tyrannosaurs would be able to run quickly and catch fleeted-footed prey, but as they got bigger, I see them hiding and quickly lunging out,' Hutchinson says. 'There's no pursuit. They slam their massive head against a large body, deliver a quick killing bite, and then back off.'

Currie welcomes theories that test aspects of his dino gangs theory because they make him ask further questions and dig deeper for answers. Each time he finds another piece of the puzzle, he steps back to re-evaluate the parts of his theory that the new information affects. In this case, Hutchinson's work adds to his belief that speedy young tyrannosaurs helped the slower adults.

'Fundamentally, this is one of the reasons people have argued that large tyrannosaurs (including *Tarbosaurus*) are probably scavengers,' Currie says. 'Basically these animals had become so large that they wouldn't have been able to move fast enough or for any sustained period of time to actually run down prey. The thought has been there that at best they're probably ambush predators.'

Donald Henderson, who is currently the Curator of Dinosaurs at the Royal Tyrrell Museum, is among those who believe that tyrannosaurs, including the smaller *Albertosaurus*, did not live or hunt in packs or gangs. 'I see them as probably solitary animals, coming together maybe in the breeding season,' Henderson says. 'A large animal like an 8- to 10-tonne tyrannosaur needs a huge amount of territory to find enough food to eat. I cannot imagine a group of them living together all the time and getting enough to eat.'

But the evidence that Currie has gathered that the dinosaurs lived in packs, such as the *Tarbosaurus* finds in the Gobi Desert and the *Albertosaurus* bone bed, changes that hunting formula.

'Suddenly you've got animals that are young and old living together and perhaps hunting together,' Currie says. 'Young animals, whether it's a young lion or a young tarbosaur, would have no trouble running something down because they were fast enough and light enough that they could sustain those speeds for longer periods of time. However, if what you're chasing happens to be a large dinosaur and what you're doing is cutting it out of a herd of duckbilled dinosaurs or horned dinosaurs, it may be far too big and far too heavy for you to take it down yourself. One strategy used by lions is for the younger ones to drive anything that they can't take down themselves back towards the large males.'

Currie believes it's possible that tarbosaurs hunted the same way. 'When the young tarbosaur moved into a herd of [large] plant-eating dinosaurs to single out prey, they couldn't kill it,' he explains. 'I think there is a good chance that what they did was drive it back to a big tarbosaur that did the killing.'

But Currie returns to the question of why so many tyrannosaurs are concentrated together at so many fossil sites in the world. 'One possibility is to investigate what happens with a modern, large, primitive and distantly related carnivore such as the Komodo dragon,' he says. 'Maybe understanding what they're doing when they're feeding will give me more insight into what happened with the *Albertosaurus* bone bed or the groups of tarbosaurs in the Gobi Desert.'

chapter 8

scavenger or predator?

There has long been a debate about whether large meat-eating dinosaurs were scavengers that ate whatever was lying around, or predators that hunted down other dinosaurs. If they were scavengers, these dinosaurs would not have had the need to pack and hunt cooperatively in gangs. But if they were predators, it is almost certain that they would have needed to cooperate to get enough food.

Currie believes speed played a factor in the hunting behaviour of large carnivorous dinosaurs. He can see that a juvenile tarbosaur could have run in the lower range of speeds possible for an ostrich, but aided by the work of John Hutchinson, he has also determined that a heavier adult *Tarbosaurus* was much slower. Some scientists believe that adult *Tyrannosaurus* and *Tarbosaurus* were too slow to hunt prey themselves, and therefore had to rely on scavenging to survive. Because *Tyrannosaurus* and *Tarbosaurus* were such large carnivores, they would have required very large home ranges to feed themselves – that is unless they engaged in some sort of cooperative hunting with their faster young playing a key role.

Jack Horner believes that *Tyrannosaurus rex* was simply part scavenger, part predator. Horner recently complete a 10 year-plus study of dinosaur remains found in the Hell Creek Formation in Montana, where the first *T. rex* was discovered. In the study, Horner counted all the specimens found there in an effort to measure the predator–prey ratio and therefore possibly determine if *T. rex* was a predator or a scavenger.

'We came to the conclusion that *T. rex* was very common,' Horner explains. '*Triceratops* was the most common and tyrannosaurs and duckbills were the second most common, which puts *T. rex* very high in numbers, much higher than a pure predator would be. This tells us that *T. rex* is most likely an opportunistic feeder that both took down prey and scavenged.'

To further delve into this question, Phil Currie wants to compare *Tarbosaurus* to one of its closest modern analogues, the Komodo dragon. These giant lizards have survived in a remote corner of world, and remained almost unchanged for nearly 5 million years. In the head, jaw and skin, they are similar-looking to large theropod dinosaurs, and there is a possibility that as apex predators they even behave the same way.

'One of the wonderful things about being a palaeontologist is that if we want to understand ancient life, we need to look at living animals that represent what dinosaurs were,' Currie says. 'While the closest living relative to a *Tarbosaurus* is a bird, the Komodo dragon is in some ways a better analogue.'

Komodo dragons are one of the fiercest predators on the planet. The only thing that can challenge their dominance is another Komodo.

'The advantage of studying Komodo dragons is that even though they are not closely related to dinosaurs, they are large with similar constraints on their bodies, and are the top predator in their environment,' he explains. 'In science, you can't come up with an absolute and say, "This is it. This is what the animals were doing." You can come out and say, "Well, it's reasonable that they were doing something between this and that."'

To study the lifestyle and feeding habits of the Komodo dragon, Currie travelled to the island of Rinca, in Komodo National Park, Indonesia in the autumn of 2009. One of the last great wildernesses left on Earth, its ecosystem bears a striking resemblance to that of the ancient Cretaceous. It was Currie's first time witnessing Komodo dragons.

Komodo National Park is located in the centre of the Indonesian archipelago between the islands of Sumbawa and Flores. The climate on the islands is pleasant because it is moderated by the sea on one side and a cold ocean current on the other. 'I'm not

The Komodo dragon, the fiercest living predator today, has been compared with the Tyrannosaurus.

sure the climate would have been all that different between the two eras – Indonesia now versus Alberta then, even though we were further north than we are now,' Currie says.

Trekking through the landscape of Rinca searching for Komodo dragons (also known as oras), it is easy to imagine that this is what parts of Asia might have looked like 70 million years ago. There are rivers and streams, semi-tropical parklands vegetation and all sorts of animals. Perhaps this is just a smaller version of dinosaur land from so long ago.

Once inside Komodo National Park, it doesn't take long to spot these incredible animals. Not far from the park's headquarters, numerous dragons are lounging around. Because most of them have recently eaten, they do not pose a threat to humans so Currie is able to move closer for a good look.

'This is probably as close conceptually to a living *Tarbosaurus* as I'll ever get,' he says, standing within 10 metres of a Komodo dragon. 'Clearly, seeing big animals like oras and how they interact is going to teach me a lot. I'll walk away with a much better understanding of *Tarbosaurus*.'

Currie finds a spot a safe distance away where he can observe the Komodo dragon and compare its anatomy with that of a *Tarbosaurus* – eyes, snout, teeth, claws, skin, tail. He recalls dinosaur skin impressions that he has examined and determines that the skin of some dinosaurs is almost identical to that of the Komodo, though he points out that the Komodo dragon's scales are actually bigger than those on *Tarbosaurus*. Seeing the musculature of the dragons' legs, he marvels that the claws must inflict considerable damage on their prey.

'It's kind of scary when you look at the heads of those things,' Currie says. 'They are much like I can imagine a big theropod looked in several ways. First of all, we always think of big toothy grins on tyrannosaurs, and yet you look at a Komodo dragon and there are almost no teeth showing. The reason is pretty simple. If you have your teeth showing all the time, your teeth are going to dry out and the chances are pretty good they are going to split. So in recent years a lot of reconstructions of tyrannosaurs don't show the teeth unless the jaws are open. To me, it makes perfect sense for a lot of reasons, the main one being that the teeth would dry out. The animals that have big, toothy smiles like crocodiles live in water.'

Komodo dragons grow to 3 metres (10 feet) long and can weigh as much as 68 kilos (150 pounds). Their short legs prevent them from running fast, and as menacing as they look in person, they aren't as imposing as a 12-metre (40-foot) dinosaur would have been at close range. Though Komodo dragons are the largest terrestrial lizard predators we have today, these clearly aren't very large animals in comparison with tyrannosaurs.

Currie notes that the oras have adaptations in the skull that are somewhat similar to those of tarbosaurs. The Komodo dragon skull has similar kinetic joints to *Tarbosaurus*, although they are not as well developed in the latter. For example, there is a joint in the middle of the lower jaw so the animal can open its jaws a little wider to ingest large pieces of their prey.

'*Tarbosaurus* has joints in its jaws that flex the same way that you see in modern lizards like the Komodo dragon,' he says. 'When tarbosaurs ate, they would also have been similar to Komodo dragons in not having teeth in the back of the jaws for grinding up food. Neither has the specialized carnassials (which are large teeth used for shearing flesh) that a lion has. Mammals are unquestionably much better than any dinosaur or any living reptile in terms of the precise action of their teeth.'

Like dinosaurs, Komodo dragons replace their teeth regularly. This process accounted for the snaggle-toothed mouth in dinosaurs. 'Komodos and dinosaurs had a way of avoiding the dentist,' Currie quips. 'Every time a tooth got to a certain age, it would drop out and be replaced with a brand new one. Komodo dragons replace their teeth at an incredible rate. There are three or four sets of teeth in every tooth position every year.'

The oras' teeth are sharp and blade-like with serrated edges – which are like miniature versions of *Tarbosaurus* teeth.

'Serrated teeth are kind of cool,' Currie says. 'Size-wise the Komodo teeth are very different than

'Komodo dragons aren't as imposing as a 12-metre (40-foot) dinosaur would have been at close range.'

a tarbosaur's but morphologically [they] are very similar. The interesting thing is nobody really has a full understanding of what serrated teeth actually mean. The assumption is that like a serrated knife, they are serrated so they can cut through flesh. But in *Tarbosaurus* the teeth are so large and thick, so the question is if having serrations is effective for cutting through flesh?'

One of the theories developed by Currie's colleague Bill Abler was that the teeth of tyrannosaurs were serrated, like those of Komodo dragons, for trapping bacteria. Then, when they bit something, the bite was septic and the prey animal

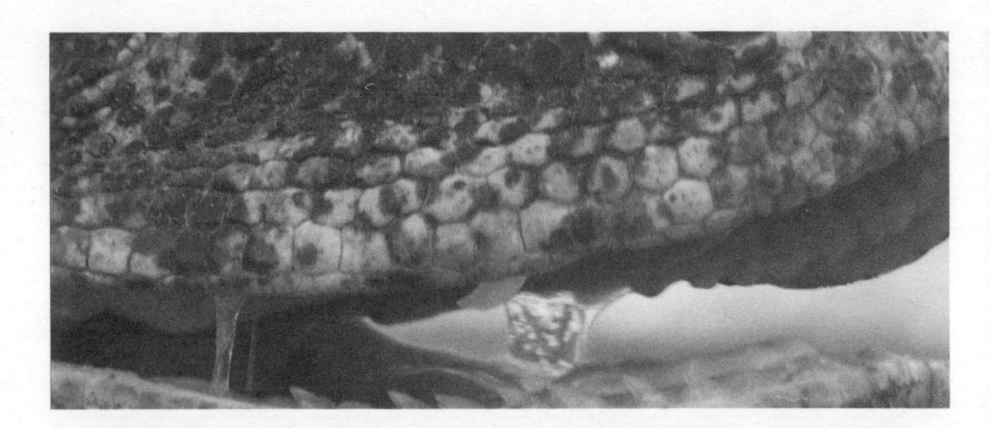

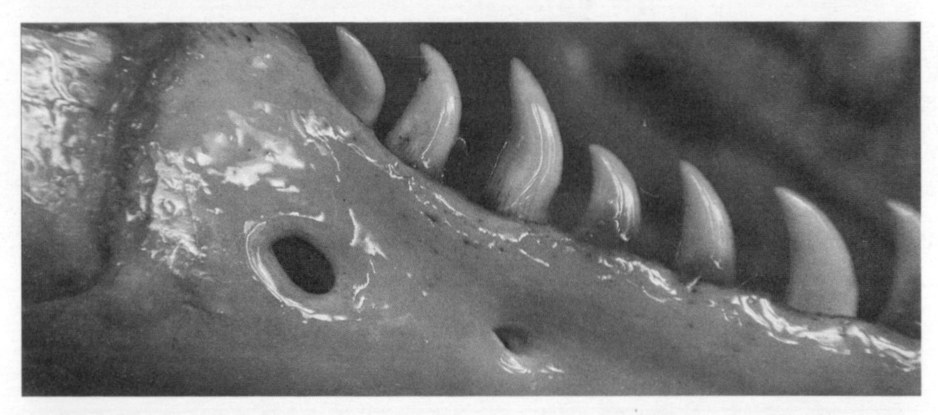

ABOVE
Phil Currie observing a Komodo dragon at very close range at the Komodo National Park in Indonesia.

LEFT
The Komodo dragon's hunting tools: poisonous saliva, serrated teeth and claws for taking down prey.

died from infection. Though it turns out that Komodo dragons have poison grooves in their teeth and not bacteria, the line of enquiry led to the conclusion that the serrated teeth in *Tarbosaurus* were used for tearing flesh and crushing bones.

'The thickness of the tooth of a *Tarbosaurus* is big because they are biting the bone as well,' Currie says. 'You can't bite into bone if you have really thin serrations because they will sheer off. So tyrannosaurs have this compromise where they have the big teeth for crushing bone but they are serrated for cutting the flesh itself. They are beautiful in cross-section. Like other theropod dinosaur teeth, if you look at two serrations side-by-side there are razor-sharp ridges between them. Furthermore, there is a slot between the razor blades. As the sharp tip of each serration is dragged across the flesh of the prey, it hooked the muscle fibres. As they passed down the slots between the serrations, the fibres were severed by the razor blades. And so the *Tarbosaurus* tooth is well adapted for breaking bones, because it's a big fat tooth with big fat serrations, but it also has those razor sharp blades between the serrations that will cut the fibres of the muscles.'

After returning from Indonesia, Currie searched for Komodo dragon teeth to study in his lab in Alberta. Despite the fact that the dragons shed their teeth frequently, swallow them and discard them in waste product, very few scientists actually collect Komodo dragon teeth. Currie finally found some at the Toronto Zoo, and they were shipped to him so his students could make more accurate comparisons with the teeth of *Tarbosaurus*.

'There's no question that having serrated teeth in a larger animal definitely increases their ability for cutting flesh,' he says. 'However, tyrannosaurs had wide teeth – kind of like big bananas – to crush bones.'

Komodo dragons are intense but somewhat lazy hunters. Though they can run very quickly over short distances, they soon tire and slow down. To help them corral prey, they often use their secret weapon, poison saliva. A Komodo dragon only has to deliver a single bite to a buffalo, deer or wild boar. Even if the animal escapes, over the next few weeks it gradually loses strength and dies. The Komodo, with a sense of smell that allows it to find a rotting animal up to 8 kilometres (5 miles) away, slowly tracks it down and gorges.

'Komodo dragons are very effective as predators, but they don't always get everything they go after,' Currie says. 'What very often happens with the Komodo dragon is that it will wound its prey, but the prey will escape. But the escape is only temporary because it was recently confirmed that they have poison glands in their gums around the bases of their teeth that cause their bites to be deadly. The bitten animal may escape, but eventually it dies and is tracked down.'

To enable Currie to study the eating process at close range, a dead boar that has already been poisoned by a Komodo is hung from a tree. Two Komodo dragons – each one almost 3 metres of muscle and aggression – smell it and arrive for dinner. Soon, Komodo dragons come from everywhere. This has turned into a feeding frenzy. With all the teeth frantically slashing and tearing, it is hard to tell one animal from the next.

Several questions emerge for watching the Komodo dragons feast. Is this how *Tarbosaurus* behaved during the late Cretaceous era? Could this be a clue to the Gobi Desert mystery of why so few prey animal skeletons have been found in the

As many as nine Komodo dragons worked together to tear the wild boar apart.

desert? Did the tarbosaurs simply eat every last piece of their prey leaving no traces of its prey to fossilize?

As the Komodo feeding frenzy continues, Currie notices that the giant lizards are devouring everything, from whole leg bones to ear cartilage. With practically the whole pig gone, one Komodo is trying to stuff what's left of the pig's head down its throat by pushing it up against a tree. The Komodo dragon's jaw is spread so wide that the bones are almost coming apart. The entire group mauling lasts just 17 minutes. Once the meal is finished, Currie moves in closer to survey the scene. He is amazed to see that every last bit of the pig has been eaten.

'If they can do that without crushing bones, imagine what tyrannosaurs did with their capabilities,' Currie marvels, trying to picture what might have happened 75 million years ago. 'I don't think any hadrosaur had a chance with any tyrannosaur eating it.'

The situation also makes him contemplate the Gobi's tarbosaur riddle. The fact that somewhere between 30 and 50 per cent of *all* dinosaur finds in the Nemegt Basin have been *Tarbosaurus* suggests that 30 to 50 per cent of *all* the animals in that Late Cretaceous ecosystem would have been the apex predator *Tarbosaurus*. However, this does not make sense.

In a normal ecosystem, the apex predator comprises about 5 per cent of the animals, thereby leaving enough prey to satisfy it. Therefore, in a normal situation – and using very simplified numbers – for every *Tarbosaurus*, scientists would expect to find 19 prey dinosaurs. But they don't. However, when the dinosaur footprints of the Gobi are studied, they confirm the expected ratio of carnivores to herbivores – only about 5 per cent of the footprints are from *Tarbosaurus* and the rest are from various herbivores. So the mystery is, why are palaeontologists finding so few skeletons of animals other than *Tarbosaurus*?

A possible explanation emerges from watching the Komodo dragons. Unlike most carnivores, when the Komodo dragons eat something, they eat *everything* – every last piece of meat, skin, hair and bone. Perhaps *Tarbosaurus* (and by extension all tyrannosaurs) ate this way. Perhaps *Tarbosaurus* devoured every last bit of its prey, just like the Komodo dragon, and there were no skeletal parts left to fossilize. This could explain why the skeletal finds in the Gobi are so out of proportion.

'That was certainly one of the thoughts that I came away with,' Currie says. 'I was amazed that there was nothing left when the oras ate. That might tell us why so few prey animal skeletons had been found and also why the number of tyrannosaurs was so high, because if the tyrannosaur ate everything, then their own proportion in the population would appear to be much higher. It's pretty cool because in a layer of footprints in the Nemegt Basin, *Tarbosaurus* makes up 5 per cent of the footprints at most. This is overlain by a layer where skeletons of *Tarbosaurus* make up 50 per cent of the specimens. This in turn is overlain by another level of footprints where *Tarbosaurus* is rare again; and so on. There is clearly some kind of bias going on with the preservation of *Tarbosaurus* in the skeletal layers.'

However, Currie points out, this ratio is unique to the Gobi and therefore other explanations are needed for other locales. 'It may not work that way in Dinosaur Park because with the big river systems, the bodies of the herbivores were picked up by the rivers before they were eaten and then they got buried in the river,' he explains.

After the feeding frenzy, some of the Komodo dragons are so full that their stomachs are completely distended. Currie thinks that the *Tarbosaurus* might have appeared this way after a big meal.

The Komodo dragon can consume a tremendous amount of meat at one sitting – up to 80 per cent of its body weight. This gives the lizard an obvious

'The Komodo dragon can consume a tremendous amount of meat at one sitting.'

advantage in the survival game because, with that kind of physiology, after gorging itself, the ora can then remain inactive for a long period of time and not worry about hunting for its next meal for a week or more.

If the same held true for dinosaurs, that means a 5-tonne *Tyrannosaurus* would have consumed an astonishing 4 tonnes of meat at a time. However, Currie does not think that the eating capacity of *Tyrannosaurus* matched that of the Komodo dragon.

'Tyrannosaurs are built a little bit differently in their body proportions,' he says. 'First of all they're bipeds. They're animals that had the main part of the body suspended over the ground with no support between the body and the head. I don't have any doubt at all that tarbosaurs probably gorged themselves when they had the opportunity, but I think their stomachs in relation to their body size were probably much smaller than what we see in Komodo dragons. Their stomachs were suspended up in the air, and they were held up by muscles and some bones in the stomach region so they couldn't do quite the same thing as oras.'

Currie also points out that the larger the animal, the more efficient it is in terms of processing food and using that food for energy to keep going over a long period of time.

'The bigger you get, the smaller your surface area is in relation to your mass or body volume, and because of that you don't use up as much energy on a day-to-day basis. That's why humans don't eat as much compared to our body weight as a mouse does compared to its body weight. Tyrannosaurs probably did eat 100 kilos [220 pounds] a day or more, but compared to its body weight of 4 to 6 tonnes, that is not a very big proportion of its body weight.'

The opportunity to observe the Komodo dragons has opened up new avenues of thought and explanation for the Gobi riddle for Currie. They are similar to tarbosaurs in some ways. They do cooperate in a way; they don't eat each other when they are gorging themselves on the same prey. However, there is a misconception that Komodo dragons are scavengers who bite an animal, wait for it to die, and then track it down and devour it.

'Komodo dragons are primarily predators. They ambush their prey by sitting beside the path. The majority of the food they eat they kill themselves and eat right away. Some people place a lot of emphasis of the fact that they bite a prey animal and then they follow it and wait until it dies. That happens, but usually it is only the really big animals like water buffalos and horses that they can't kill immediately. If they can wound these very large animals, then eventually they will die because of the poison. That is a sort of scavenging. But 80 per cent of the time, what they catch, they kill and eat right away. And if they have a choice between a dead animal here and a live animal there, they'll go for the live animal.'

Another thing that is clear is that a single Komodo dragon cannot take down a larger animal like a water buffalo by itself – the same way a small *Tarbosaurus* couldn't take down a large sauropod. Oras have another way of getting the water buffalo – their poisonous, serrated teeth. But how did *Tarbosaurus* do it? By hunting in gangs?

'There are a lot of clues here as to what happened 75 million years ago with dinosaurs,' Currie says. 'All we have to do is look at the modern world and it gives us the best evidence we have to understand what happened in the past. But now I have to take this a little further and look at the possibility that *Tarbosaurs* did hunt cooperatively.'

* * *

Back at the Tswalu Kalahari game reserve in South Africa, Currie studies the greatest modern social hunters: lions. By observing them in the wild he hopes to see how they interact, how their social structure works, and how they work together to hunt in gangs.

The present-day African savannah provides us a template of how the Late Cretaceous ecosystem may have worked. There are large numbers of prey animals, far outstripping the numbers of predators, and among the predators there is one at the very top of the food chain – the apex predator. Seventy million years ago the prey animals were hadrosaurs (*Saurolophus*), sauropods and ankylosaurs; now in Africa they are antelope, giraffe and zebra. In the prehistoric Gobi, the apex predator was *Tarbosaurus*; today the undisputed king of the grasslands is the lion.

Seeing the world's most iconic apex predator is a life-affirming moment for Currie. Though he's slightly awed by the adult male lions, which are 225 kilos (500 pounds) of muscle with sharp claws and vice-like jaws equipped with deadly teeth, he quickly remembers that this modern apex predator would have been dwarfed by a 12-metre (40-foot) long, 5-tonne plus *Tarbosaurus*.

Living in gangs requires a high degree of socialization. There are dominant members of the lion pride mixed in among the sub-adults; there are fights, squabbles over hierarchy; and they care for young and sick individuals – all of which are aspects of socialization that Currie thinks could have been present 70 million years ago.

'We have different lines of evidence for a lot of those things,' he explains. 'So for example, when you look at the skulls of tyrannosaurs, very often they have healed bite marks on their jaws, around the eyes and on their snouts. The bite marks are from other tyrannosaurs. The fact that they are concentrated on the face is a strong suggestion of socialization. It has got nothing to do with actual fighting. It could be mating, because there are a lot of animals today in which the male bites the female's head and holds on during the mating ritual. The fact that there are a lot of bite marks on the faces of tyrannosaurs of all ages suggests that it may not have had anything to do with mating, however.'

A pride of lions sharing the spoils of the hunt in the African savannah.

Lions are among the most social of all hunters. As mammals, they take care of their young for up to two years, which results in long-term bonds being forged at an early age. They undergo a learning process that teaches them essential survival skills, and by the time the lions are fully grown, they are part of a tight, cohesive unit – as well as some of the best hunters in the world.

'There's no question at all with social predators one of the biggest advantages is not just clustering together to hunt; it's the fact that the young learn from adults so they have a much better chance of surviving than other species of predators that don't have this kind of social behaviour,' Currie says.

Once the heat of the day gives way to the cool evening air, what seemed like a group of cuddling cats transforms into a gang of silent killers. Though the conventional wisdom is that lions hunt large animals, when it comes to an evening meal just about anything will do.

Curries watches the lions hunt a porcupine. At first glance, it doesn't seem like a big deal for a huge lion to kill a relatively small porcupine, but the fact of the matter is that one lion, no matter how large, cannot take down a porcupine on its own. The porcupines use their sharp quills to defend themselves, much like an armoured dinosaur might have done against a *Tyrannosaurus* millions of years ago. So even with such a small creature, it takes a group of lions to turn this prickly problem into a meal. Two approach from one side, while another grabs the porcupine from behind – they cooperate.

The young lions learn to hunt at their mother's side, and they also practise their hunting skills. Cooperation implies intelligence, but it also requires lots of practice. Currie witnesses a group of young lions catch sight of a group of antelope, quietly disperse to surround the prey, and attempt to get close enough to be able to surprise the potential kill; Each lion seems to know its role, and it becomes clear how important their intelligence is. The young lions try – but fail – to catch the antelope.

Once the real hunt is on at night, things change. One female lion takes the lead, tracking the target, and then another swoops in from the other side. Excellent night vision allows lions to pick out their

'A group of cuddling cats transforms into a gang of silent killers.'

prey in the dark. The gang members spread out left and right, spooking the antelope herd, with the goal of sending them right into the jaws of the more experienced lions lying in wait.

'It was amazing to watch because it was so instinctual in a way,' Currie enthuses. 'There were as many as 15 or 20 antelope at a distance because they don't let the lions get too close and they are on the lookout all the time. When the lions are hunting they are constantly sniffing the air. Then whatever lion picks up a scent first, he or she stops, and when she or he stops, the other ones stop. There is no verbal communication between them, but they all know their roles. They split up and move out. They will all be watching for a long time. What they want to do is get close enough and if they can, they will surround their potential prey. When one starts to rush, then they will all come out of hiding and close in.'

Currie observes that the lions hunt differently than do Komodo dragons. 'The Komodo dragons certainly were eating cooperatively in terms of being able to pull something apart by working

'Tyrannosaurs had strategies that worked, because they were incredibly successful animals.'

together, but these guys are doing more than that – they're actually thinking together,' he says. 'They're communicating in some way. They know what the objective is, and they know that they have to try different ways to get it. There's a different level of intelligence in this kind of hunting and it's hard to imagine that tyrannosaurs could have done anything like this; probably tyrannosaur hunting behaviour was somewhere between these guys and the Komodo dragons.'

Exactly where tyrannosaurs fit between these two modern-day animals comes down to intelligence.

The Komodo dragons' teeth, jaws, skin and claws are similar to those of tyrannosaurs, and the two animals probably devoured prey in a similar fashion. But though Komodo dragons act cooperatively, they hunt as individuals. Clearly they are not smart hunters, nor do they have to be because they rule the landscape. Currie now must delve into whether tyrannosaurs were smart enough, and had advanced-enough senses, to hunt cooperatively.

'I think tyrannosaurs, whether or not they actually hunted [in] the same kind of cooperative way as lions, probably had strategies that worked, because they were incredibly successful animals for most of the Late Cretaceous,' he says. 'There is no question at all that what they were doing worked. Over that period of time, they had lots of time to develop their instincts for cooperative hunting at some level.'

chapter 9

dinosaur intelligence

ow smart were tyrannosaurs? It is one of the toughest questions to answer and one of the most critical to Phil Currie's dino gang theory. Currie believes that group behaviour in top predators comes in part from a high degree of intelligence. Determining the intelligence of living animals has proved difficult enough for scientists, so calculating it for extinct animals presents an even greater challenge.

In the past, scientists would actually run a saw through the dinosaur skull to gain access to the inside, which ended up destroying the specimen. But modern technology has been a great help to palaeontologists in this area by allowing them to examine fossilized dinosaur skulls with sophisticated CT scans and to generate computerized models of a dinosaur's brain.

'What CT scanning has done is open up whole new vistas because it has allowed us, without actually damaging any kind of priceless fossil, to

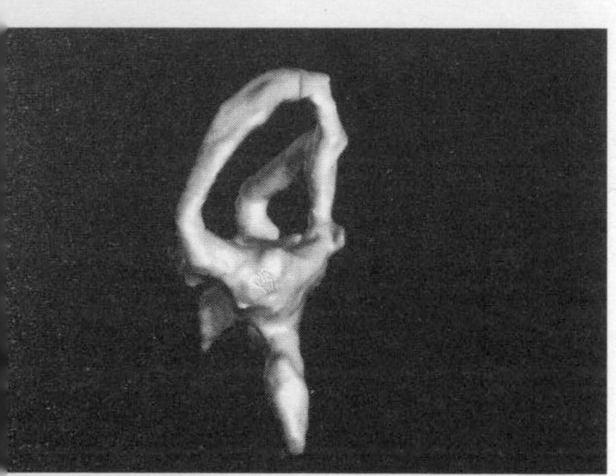

peer inside,' explains anatomist and dinosaur brain expert Larry Witmer. 'The X-rays look through the bone [and] rock that entombs the fossil allowing us to see inside. What we're looking for are signs of tissues that once were there. Within the skull are various spaces where the brain once was and where nerves, blood vessels and air sinuses once were.'

When CT scans on dinosaurs were first done some 20 years ago, they were very primitive. In some cases, the bone was so thick the scientists couldn't see anything inside. In other cases, the images were so fuzzy that it was hard to draw any conclusions. Modern CT scans are far clearer, and show details of the anatomy as fine as the semicircular canals in the inner ear. The digital images can also be used for other types of studies. For example, scientists can use something called finite element analysis on the digitized skull models, and this engineering program tells them the strengths and weaknesses, plus the stresses and actions on certain parts of the skull. That provides them with information about how the animals developed. They theorize, for example,

TOP LEFT
An endocast of a braincase.

that a bone wouldn't have strengthened in a certain direction unless the animal's muscles were pulling in that direction, indicating that the animal needed some kind of resistance against the stresses that were introduced by its huge bite.

'All of these software analysis programs really help us to visualize and look at the maximum/minimum possibilities,' Currie says. 'If you make an assumption that tyrannosaurs were using their jaws a certain way and do CT scans, the resulting models and analysis might tell you that the jaw would break if they tried to bite that way. That would tell you that you are wrong in your assumptions. This kind of study does not give you absolute answers, but it does give you a range of possibilities so you can constrain your assumptions much better.'

To better understand how this ancient killer might have behaved, Currie travels to the idyllic campus of Ohio University in Athens, Ohio to work with Larry Witmer analysing the dinosaur skull and brain. Currie has known Witmer for years. He met him in Poland in 1981 when Witmer was still a student and had just completed a controversial study on the structures of sinus systems of dinosaurs, in which Witmer reached the conclusion that there were many similarities between birds and dinosaurs. Currie laughs at the irony. 'His supervisor was one of the people who doesn't believe that birds came from dinosaurs,' he explains.

LEFT MIDDLE AND BOTTOM
Digital reconstructions of the semicircular canals in the inner ear.

Breakthroughs in laboratory technology have opened up new possibilities into the study of dinosaur behaviour. 'We used to think that some things were out of reach,' Witmer says. 'Behaviour was just too much speculation. However, we have some new ways of looking at it. We never used to think that we could ever come up with the behaviour of an animal from 70 million years ago. We are opening up new vistas for reconstructing the behaviour of extinct animals. We can now reconstruct what the bones are telling us about behaviour. We can bring them back to life. What's remarkable is how rapidly our views are changing about tyrannosaurs.'

Witmer spends much of his time figuring out what went on inside dinosaurs' heads many millions of years ago, or as he puts it, trying to understand the 'functional morphology of the heads of vertebrates'. Did they merely scavenge for food, or did they use their brains to cooperate, which would have allowed them to hunt down prey in organized gangs?

'A solitary hunter doesn't require a complex thought process; it runs down its meal and gets it,' Witmer says. 'Being a communal hunter, though, requires more sophistication in its cognitive abilities because it has to know that it's someone else that's going to be doing the killing and it is maybe playing a different role.'

One of the things that Witmer can determine using CT scanning is the size of the fossil's brain. He can then scale it up or down from a modern animal to draw a conclusion on its relative size. A *Tyrannosaurus rex* skull he examined was an incredible 1.5 metres (5 feet) in length.

'If we think about taking a reptile like a lizard and scaling it up to the size a *T. rex*, the brain of *T. rex* would actually be about three times the size of the brain of that lizard,' Witmer says. 'So even though the brain seems pretty small, it's actually relatively expanded. And what's key is there are certain parts of it that are even more expanded. The higher centres of the brain, the cerebrum, really seem like they're even extra expanded.'

Brain size is one way intelligence can be estimated. *Troodon*, a relatively small theropod that

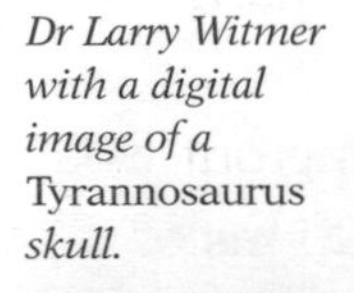

Dr Larry Witmer with a digital image of a Tyrannosaurus *skull.*

was 2.4 metres (8 feet) long and just under a metre (3 feet) high, had the largest relative brain size of any dinosaur. Sauropods had the smallest, but that does not directly correspond to intelligence. There is trackway and bone bed evidence suggesting that sauropods were capable of complex behaviour.

'We tend to think that animals with the big brains are the smartest and ones with the little brains are the dumbest, and the reality is our understanding of modern behaviour says that this is not necessarily the correlation,' Currie explains. 'The brain size doesn't necessarily tell you the whole picture.'

Though a sauropod was physically larger than a tyrannosaur, it had a smaller brain. Given that sauropods had a complex social structure, this raises the possibility that tyrannosaurs may have been more complex socially than sauropods, or at least as complex.

'The sauropods, even with their little, tiny brains, definitely had complex social structures,' Currie says. 'And we can see that in the trackway sites where they are interacting with each other and they are moving in herds. So if these guys could do it, the potential has to be in tyrannosaurs too.'

According to Witmer, the sauropod brain is about half the size that we would expect for such an animal, whereas the tyrannosaur's was about three times as large as expected.

'We do have evidence that these guys had some kind of fairly sophisticated social behaviour, which raises the question: how much brain power does it take to have some fairly complex social interactions?' he says. 'What it also does point out is that the size, the expansion of what we see on the cerebrum in *T. rex*, suggests that indeed it probably did have the cognitive capabilities to pull off some fairly subtle and maybe even impressive social behaviours – potentially even communal hunting. Certainly what we can say is that *T. rex* had the brain power to outwit most of the prey animals in its environment.'

One scientific way to find out whether tyrannosaurs were scavengers or predators is by seeing if their senses were developed for hunting. This means examining the skull and recreating a tyrannosaur's brain so the scientists can examine the inside of it.

Witmer has a juvenile *Tarbosaurus* skull that measures 770 millimetres (30 inches) long. The braincase is surprisingly roomy. For its time, *Tarbosaurus* had a big brain; indeed, even compared to many modern animals *Tarbosaurus* had a big brain. Looking at the skull, Currie can see that the dinosaur's eyes are slightly apart, giving it perfect binocular vision – and a predator's gaze.

However, Witmer's methods go much deeper and aim to find out exactly *how* this animal could see, hear and smell, and ultimately, how this would have affected its behaviour. The first step is to take the skull of the bones to the university hospital and give it a CT scan.

CT scans are commonly used in hospitals to see inside patients. They use X-rays to produce 'slices' of computer images through the skin so doctors can see what's going on. That same technology can be used on extinct animals like dinosaurs. However, when scientists scan a fossil of a long-dead animal, there is no risk of damage to living organs so they can increase the power of the X-rays to see more detail.

The CT scan of the *Tarbosaurus* skull will allow Witmer to use advanced visualization software to produce a complete 3-D image of the cavity where the brain would have been. The shapes of the sinuses and airways can also be seen in other parts of the skull, and provide information on other sensory systems.

'When we get a fossil or a CT scan of a fossil, the natural thing to look at is the bones,' he explains. 'That's what time has left us. The frustrating thing is that the bones themselves don't do anything. What time has done is strip away all of the tissues that clothe and animate the skeleton – the blood vessels, the muscles, the nerves, the brain tissue, the air sinuses. These are the things that make up and cover the bones and make it into a biological organism. So in a sense, although we are indeed interested in the bones, we're really interested in all these things that are in the spaces.'

Some of the most important scientific data comes from the empty spaces in the skull. It is Witmer's job to figure out what was in them. When he finds a canal, he must determine whether it was filled with a nerve or a blood vessel. As he delves into the cavities of the skull that were occupied by the brain and the surrounding blood vessels, he can get a glimpse of the brain structure.

Currie is hoping that studying the inside of the *Tarbosaurus* head will confirm that these beasts weren't dull-witted scavengers, but rather that they

'We're seeing things that nobody has ever seen.'

were capable of more complex behaviour, like hunting in a pack. For Witmer, it is a chance to help Currie test this theory using brain science. 'In some respects this is almost the same level of excitement as a field palaeontologist experiences as he starts to brush away the sand and dirt that's covered the fossils for millions of years,' he says. 'We're seeing things that nobody has ever seen. In some cases we're really confused because nobody really knows what it's supposed to look like; it's brand new anatomy.'

The CT scan takes place in a CT scanning facility located in O'Bleness Memorial Hospital on the Ohio University campus. Witmer and Currie carefully place the skull onto a CT scanning bed. They move into an adjacent control room to shield themselves from the intense radiation. As Witmer operates the CT scanner, a series of images that are slices of the skull come up on a screen. As the scan probes the cranium's nooks and crannies millimetre by millimetre, the long-extinct animal appears to come to life. The scans clearly show the size and outline of every region of the endocast, which will allow him to digitally reconstruct the brain.

After the CT scan is complete, the data is taken to Witmer's lab and run through high-powered computers and painstakingly evaluated. From the images, he can determine what the brain looked like and build a 3-D computer model of it.

It is evident from the skull that the eyes were large. 'If you have large eyes, you have a lot of [light] receptors, and therefore you probably have a good sense of sight,' Currie says. 'Also, if the optic lobes are well developed, then the animal probably had a good sense of sight.'

For more detail, Currie relies on Witmer's brain model. This reveals some fascinating details about the tyrannosaurs' senses of hearing, vision and smell. The size of the cerebrum, the main part of the brain, allows Witmer to estimate how advanced the dinosaur's behaviour might have been. Reviewing each area in detail paints a clearer picture of how sophisticated tyrannosaurs were.

To assess the animal's vision, he examines the optic nerve, which carries visual information from the eyes. The relative size of this optic nerve tells him something about the amount of information that can be carried.

'In tyrannosaurs, the optic nerve was a very large structure that tells us that vision was important,' he says. 'The collecting of information from the eyes and retina was important to these animals. It turns out that there is another element to this structure that also tells us something about vision.'

He focuses on the inner ears. The inner ear has two major compartments, an upper portion that relates to the sense of balance and a lower portion that relates to hearing. He lays out in detail how the upper portion suggests that tyrannosaurs had sophisticated vision.

'It winds up that predators have a very highly tuned system that allows them in a sense to keep their eyes focused on their prey as the predator is moving,' he explains. 'How this system works is as a predator moves and turns its head while it's

CT scanning probes the nooks and crannies of a tyrannosaur skull.

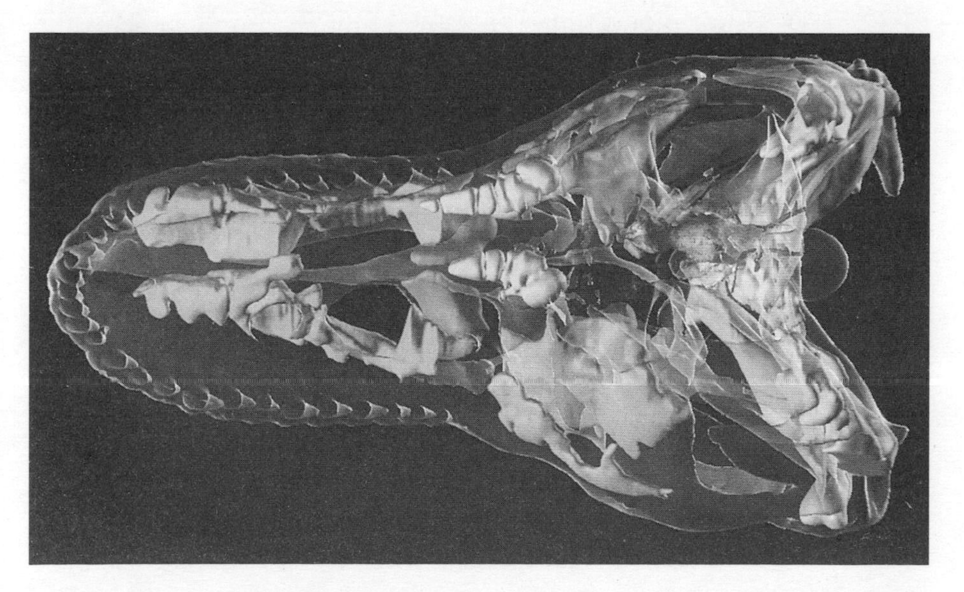

watching prey, these delicate canals actually sense that turning movement of the head and space. These canals then send that information on head position to the brain, which then in turn sends that information to the eye muscles, which then moves the eyes around in space. So as the animal is moving and turning its head, it's actually allowing the eye muscles to move in a compensating way so that it keeps the prey firmly within its cross hairs. We expect these kinds of mechanisms to be highly tuned in a predatory animal. When we look at a tyrannosaur what we see are the long delicate canals that we expect to find in an animal that has these mechanisms to stabilize gaze to keep the prey fixed and in focus within its gaze.'

Witmer concludes that tyrannosaurs had a high degree of advanced 'gaze stabilization', meaning that they would have been able to run fast and still see clearly. The stability of gaze means that a tyrannosaur could keep its eyes on running prey as it chased them through the ancient environments.

'If you think about focusing your eyes on the tip of your finger and you shake your finger, the image blurs,' he explains. 'But if you focus your head on the tip of your finger and turn your head around, the imagine stays fixed and in focus. What that's reflecting is this mechanism of the inner ears. As your head turns, the inner ears sense that and communicate that through the brain to the eye muscles so that the eye muscles are making these minute rapid adjustments of the position of the eye such that the image stays fixed and in focus. What we see in predatory animals and what we see in tyrannosaurs is a highly sophisticated development of that system.'

Depth perception is also critical for predators. When a predator is striking at something, if it doesn't have a good depth of field, it can miss very easily. A herbivorous dinosaur was unable to gage depth of field as well as a carnivorous dinosaur because each of its eyes faced more to the side. This gave it a wider field of vision so that it had a better chance of seeing an approaching predator before it got too close. Most herbivores had a division between their eyes, which faced sideways. The advantage of that is they did not care so much about their fine range vision; rather their eyes were spread wide so they see if a predator was approaching from either side.

'If you close your eyes and know where your food is – in this case that the grass on the ground is x number of inches above your feet – then you don't need to see it; you can still find it,' Currie says. 'But if you are worried about maximizing your chances of seeing a predator and having 180 degrees of vision, it's going to work for you much better.'

The depth perception in tyrannosaurs suggests they had advanced vision. 'When you look at their eyes, their field of vision is in front,' Currie says. 'One of the most amazing displays to me is "Sue", the *Tyrannosaurus rex* at the Field Museum in Chicago. If you stand in front of that specimen, what you will see is that the two eye sockets are facing you. That animal had 3-D depth perception. There is no question at all. It's not like *Jurassic Park* where the animal couldn't see you if you didn't move. I don't know of any carnivores that are like that. Basically, tyrannosaurs and other carnivorous dinosaurs have their eyes set up that way because they have very good depth perception.'

The CT scan is also able to determine the size and shape of the semicircular canals in the inner ear, which affected the agility and balance of the animal. Analysing the inner ear, Witmer concludes that tyrannosaurs were highly agile and as a result

'Tyrannosaurs were probably animals that were fairly jumpy and had relatively quick movements.'

had the ability for quick head and coordinated body movements. He is almost certain that the inner ear canal demonstrates that they had good control of their ability to move their heads from side to side.

'One thing that we would expect any predator to have would be some measure of agility – the ability to catch an animal [that] does not want to be caught. So we can look at the brain sensory system to see if we can get any clues on that. It turns out that the back part of the brain houses the part of the brain called the cerebellum, which controls movements and motor coordination. It winds up that the cerebellum in tyrannosaurs is a pretty large complex portion of the brain, suggesting that the agility in motor coordination was very important. Also we can see the portion of the inner ear, which is associated with balance and the sense of equilibrium. These canals of the inner ears are very long and delicate, suggesting that the sense of balance and equilibrium was very highly developed in tyrannosaurs. These are the kinds of attributes that we would expect to find in a very agile predator, and that's in fact what we see.'

Witmer concludes that this information suggests that tyrannosaurs were not slow-moving, lumbering animals; rather they were probably animals that were fairly jumpy and had relatively quick movements. Certainly the younger animals would have been highly agile, but even the adults would have had relatively snappy movements of the eyes, head and neck. The foundation for this kind of behaviour can be seen in the brain and the inner ear.

There are also sinuses associated with the inner ear that dampen the air behind the eardrum so that certain frequency sounds can be more easily picked up, mainly low frequency sounds. 'By looking at the sinuses, by looking at the auditory region of the ear, and by looking at the semicircular canals, we can get a sense of whether or not they heard well,' Currie says. 'Tyrannosaurs are pretty amazing, including *Tarbosaurus*, in the sense that they also had connections from the left side to the right side of the skull and vice versa. We know that modern birds have good discrimination of where sounds are coming from because, of course, a sound usually hits the left and right ears at different times (depending on which direction it is coming from), so the brain processes them differently.'

Witmer determines that all tyrannosaurs had special sensitivity to low frequencies. This would have helped them hear large herds of dinosaurs from far away, thus allowing them to begin a hunt even before they could see their prey.

'The advantage of low frequency hearing is that low frequency sounds travel over very long distances and even over rough terrain or densely forested terrain,' he says. 'What that means for a predator is that it doesn't necessarily have to see its prey. It can track the movements of herds that are at a long distance or that are shielded by vegetation. What we see in tyrannosaurs is that they had very acute low frequency hearing, [which] is another of the predatory adaptations that tyrannosaurs had.'

An acute sense of smell can also be extremely helpful in hunting. Did the dinosaurs have an extraordinary sense of smell similar to that of a Komodo dragon? By examining the olfactory lobe of the brain, Witmer shows that their sense of smell was indeed very good.

'One thing that was remarkable about the brain structure of tyrannosaurs was the size of the olfactory bulb. The olfactory bulb is the part of the brain that is associated with processing smells in the environment. We would expect predators to have a very highly attuned sense of smell simply to be tracking the movements of their prey. Likewise, what we might expect is that tyrannosaurs would have had a good sense of smell potentially for other behavioural reasons. Maybe these were territorial animals using the sense of smell to track movements of members of the opposite sex during the breeding season. One thing that's clear is that tyrannosaurs had the sense of smell that we would expect to find in a predator that was tracking prey.'

Larry Witmer's CT scans have revealed that tyrannosaurs had senses that were specifically used for hunting. A large cerebellum suggests they had mobility and motor control. The ability to hear low frequency sounds gave them a long-distance prey-tracking mechanism. A large olfactory bulb indicates that they had a very good sense of smell. Their eye sockets faced forwards, giving them good binocular vision. This is a trait in modern animals found mainly in predators. If tyrannosaurs had been pure scavengers, they wouldn't have needed the accurate depth perception that this kind of vision provides. Combined, this evidence might help put an end to the predator vs. scavenger debate.

'Looking at CT scans of the brain of any of the other tyrannosaurs, including *Tarbosaurus*, reconfirms in my mind that these things are almost certainly predatory animals,' Currie says. 'The brains are much larger than one would expect to see in a big scavenger, and they outclass any of the herbivores that they would have potentially gone after.'

Currie must now focus on how these senses affected behaviour. Specifically, he wants to know if studying areas of the brain allows something to be said about the social behaviour of these prehistoric animals.

'I wish what we could do is point to a little part of the brain and say, "That's the communal hunting lobe of the brain,"' Witmer says. 'Sadly we don't really have that [ability]. In fact, even in the modern realm we're not that fortunate. So what we can do in the fossil realm is look at the relative size of the brain parts. For example, we can see that the cerebrum is relatively quite expanded in tyrannosaurs. That does open up the possibility that they potentially were engaging in fairly sophisticated behaviours and that they may have had the cognitive power to have been communal hunters.'

Currie prefers to reverse the line of enquiry: to ask whether there's anything in the brain to say that tyrannosaurs could *not* have complex social behaviour.

Witmer agrees there is nothing to say they did not behave socially, and he goes even further. 'It would be wrong to suggest that these animals didn't have that power. I think we can say that these animals had the expansion of the right parts of the brain to make the reasonable assumption that they were pretty complicated and sophisticated animals.'

'Tyrannosaurs had the senses of efficient predators and the brain capacity to be able to hunt in deadly organized gangs.'

What the CT scan and Witmer's analysis has revealed lends support to Currie's theories about all tyrannosaurs, including *Tarbosaurus*: that they had the highly developed senses of efficient predators and the brain capacity to be able to hunt in deadly organized gangs.

'The idea of pack-hunting *Tarbosaurus* is tremendously seductive because it really changes everything about how we view the interactions of animals millions of years ago,' Witmer observes. 'What we sort of thought is that these prey animals would have been worried about some rogue tyrannosaur in their environment; now they're worrying about these roaming packs of tyrannosaurs actively acting cooperatively and intelligently. It's beyond what we would have thought before.'

chapter 10

the dino gang theory comes to life

Palaeontology is a living, breathing discipline. There are few absolutes, and theories are moulded and altered by new discoveries. For Phil Currie, exploring whether dinosaurs lived and hunted cooperatively in gangs has been a decades-long pursuit that will continue as new evidence is found and evaluated. His own quest to study dinosaur science from all areas using the past and the present as a way to understand the past has drawn out his dino gangs theory and led to an amazing conclusion that he is ready to have debated.

'The more I look into modern animals and see the range of behaviours that are possible and how that influences the shapes of the bodies, the more I'm convinced that ancient animals were doing similar kinds of things,' he says. 'So social behaviour in tyrannosaurs is not as outrageous as we would have thought 20 years ago.'

To solidify and enhance his theory, Currie has repeatedly returned to the Gobi Desert to re-examine the *Tarbosaurus* finds, as well as the *Albertosaurus* bone bed in Alberta. He travelled to the Hwaseong Dinosaur Lab in Korea to review the bones from the Gobi expedition finds of the previous four seasons. In Indonesia and Africa, he studied the modern counterparts to dinosaurs for clues – the Komodo dragon for cooperative eating habits, the ostrich for speed, and the lion for pack hunting. He then employed progressive CT scanning methods to examine a *Tarbosaurus* skull and build a computerized model of its brain.

A cornerstone of the new science of dinosaurs is the use by palaeontologists of modern animals, comparing aspects of their behaviour to that of dinosaurs, and then finding a range of probabilities for how dinosaurs lived. The flexibility to accept new lines of thought about dinosaur behaviour based on modern analogies has created new avenues of possibilities for palaeontologists assembling a picture of the 70-million-year-old animals.

'There's no question that after 35 years of studying dinosaurs, you still have things to learn, and being able to look at modern apex predators has changed the way I think about tyrannosaurs and how they interacted with each other,' Currie says. 'We originally went into this whole question about dinosaur behaviour with no hope of being able to say anything that was concrete. Yet as time goes on, by studying modern predators we come up with things that we didn't think that we would ever be able to say about those ancient predators. People are starting to accept this on a broader basis. So it's always a process where you're learning, and I'm happy to say that I've learned a lot in the last year about lions, Komodo dragons, ostriches and how they interact with their ecosystems. And it has taught me a lot and changed my thinking about how tyrannosaurs may have interacted with their ecosystems.'

Currie is now ready to take all the information he has gathered, apply it to his years of fieldwork and research, and put it out for experts and lay people alike to debate. The evidence points to the conclusion that *Tarbosaurus* and its tyrannosaur cousins lived in groups (at least at certain times of the year), that they were predators not scavengers, that they engaged in complex social behaviour, and that they probably hunted cooperatively in packs or gangs.

To evaluate the information, Currie returns to his office at the University of Alberta in Edmonton. Located within walking distance of his house, his office is on the fourth floor of the Biological Sciences Building. The academically oriented, two-room suite is filled with a mixture of hands-on palaeontology and high-tech science. Fossilized dinosaur bones from his field finds are on display. The office has separate work areas, one for his wife and fellow palaeontologist, Eva Koppelhus, to assimilate the vast amounts of information, and another for graduate students to work on computerized models. Currie's own workspace has his main computer and the laptop he carries with him in the field.

Currie has methodically put together the building blocks of his dino gang theory. He has found convincing evidence that large carnivorous dinosaurs lived together on two continents – *Albertosaurus* in North America and *Tarbosaurus* in the Gobi Desert. In 1997, he rediscovered the most important *Albertosaurus* bone bed in the world on the banks of the Red Deer River in the badlands of Alberta, which had first been found by dinosaur hunter Barnum Brown in 1910. After years of excavations, Currie has concluded that the bone bed contains the remains of at least 12 and as many as 26 *Albertosaurus*.

'Once we knew that *Albertosaurus*, which is a smaller relative of *Tarbosaurus*, was congregated in packs, then our eyes were opened up to the possibility that *Tarbosaurus* might also do the same thing.'

On several trips over a number of years, Currie identified *Tarbosaurus* finds in the Nemegt Basin of the Gobi Desert, and he visited each site to examine the evidence and determine when they died. By the end of the 2010 field season, he had identified and seen 90 sites. Much of this work in the Gobi was done over four years as part of the Korea-Mongolia International Dinosaur Project.

'We did see that large numbers of *Tarbosaurus* were being found in exactly the same levels in the

Nemegt Basin, and then in a lot of cases these animals seemed to be coming from exactly the same levels within those levels,' Currie says. 'So in other words, you could look at the bed of an ancient stream and you could see that there's a *Tarbosaurus* and a 100 metres further on there's another *Tarbosaurus* skeleton. Because they're at exactly the same level in the stream channel, then one can't help but think that they died at around the same time, in roughly the same place.'

Because Currie has concluded that these carnivores died together, he believes that they also lived together. If they lived together, he says it follows that they would have had to hunt cooperatively because there would not have been enough food for so many scavengers living in such close proximity.

'We can say that some of the animals definitely died and got buried at the same time. Even though they're at exactly the same level and within your field of vision, we can't be as sure whether they died within days or within weeks. But still we're talking about a relatively short period of time and the fact that you have a single species of animal in such a concentration tells you that there is something going on.'

Currie first wrote about carnivorous dinosaurs roaming in packs in a 2000 scientific paper entitled 'Possible Evidence of Gregarious Behavior in Tyrannosaurids'. The paper was based largely on his study of the rediscovered *Albertosaurus* bone bed.

Through further fieldwork, he has confirmed and advanced the major conclusions in the paper.

'Since the first paper was published in 2000 suggesting that *Albertosaurus* was a packing animal, we've moved far ahead,' he says. 'We now know that in South America we have a bone bed where we have multiple individuals – nine carcharodontosaurids called *Mapusaurus* – and the evidence strongly suggests it was a packing animal. We have a group of *Daspletosaurus* in Montana, which is potentially showing again a pack of animals that was going after herbivorous dinosaurs. There is the evidence in Mongolia where *Tarbosaurus* continues to accumulate, and it's perhaps explained by the fact that they were packing as well. That's not an explanation we had before, even though we knew there were large numbers of tarbosaurs there. So with a different spin on how we approached the *Tarbosaurus* problem, we are getting a better understanding. It may boil down to the fact that *Tarbosaurus*, like other tyrannosaurs, also moved in big groups.'

Currie has focused on *Tarbosaurus*, the lesser-known cousin of *Tyrannosaurus rex* that lived in the Gobi Desert. *Tarbosaurus* was a bit smaller than *Tyrannosaurus rex*, the largest carnivorous dinosaur, but not by much. An adult *T. rex* could reach nearly 13 metres (42 feet) long and weigh more than 7 tonnes, whereas the biggest *Tarbosaurus* found so far was about 12 metres (39 feet) long and weighed an estimated 6 tonnes.

'*Tarbosaurus* is pretty close to *T. rex*, but it was not ancestral because the lines split,' Currie explains. 'There was an Asian line and a North American line. *Tyrannosaurus rex* is the top of the North American line; *Tarbosaurus* is the top of the Asian lineage.'

Slightly bigger than *Tarbosaurus*, *Tyrannosaurus rex* also had slightly more advanced features. 'Because *T. rex* is bigger and had the jaw muscles that were more powerful, and because it was that much wider at the back of the skull because of the jaw muscles, that means that its eyes were facing that much further forward,' he explains.

However, Currie says that *Tarbosaurus* was more advanced than *T. rex* in the development of its arms, which he believes were used only during mating. *Tarbosaurus* had much shorter arms than *Tyrannosaurus rex* arms, or even than *Albertosaurus*, which is on the North American line of tyrannosaurs.

'These things are big seesaws. They have a lot of weight, but the front part of the body is thicker. As you get bigger you are putting a disproportionate amount of weight on the front part of your body so your tail doesn't balance the front any more. So you have to get rid of excess weight in some way to make the front part of your body shorter and lighter than your counterbalancing tail. All vertebrae in the front part of a tyrannosaur body became shorter to pull the head back toward the centre of gravity, just in front of the hips. Then many of the skull bones and vertebrae became hollow and air-filled to lighten the front part of the skeleton further. It is a beautifully engineered structure. Because *T. rex* did not generally use its arms because it was using its teeth and jaws to kill things, then it was able to shorten its arms. But it kept its arms for something else; at this stage we do not know what that function was, but the arms were still powerful even though they were short. In *Tarbosaurus*, the arms are a lot shorter.'

In appearance, *Tarbosaurus* was not the big, scaly monster that has been portrayed in movies. Palaeontologists have recovered the skin of *Tarbosaurus*. It has a very lightly pebbled texture and looks a little bit like the hide on an elephant or a rhinoceros. They have also found stomach contents that lead them to suspect that these animals consumed the meat and bones of their prey, and in the case of sauropods they also ate the internal guts. This is evident because they managed to ingest the stomach stones or gastroliths that were in the guts of sauropod plant-eating dinosaurs.

Tarbosaurus was clearly a fierce predator. It had massive jaws that allowed it to bite very forcibly. It also had sharp-tipped, serrated teeth for tearing flesh, but the thick, banana-shaped teeth also allowed them to bite through bones. Like the Komodo dragons that Currie observed in Indonesia, tarbosaurs probably devoured every last bit of their prey.

This parallel between the eating habits of Komodo dragons and tarbosaurs may provide an answer to the Gobi riddle of why somewhere between 30 and 50 per cent of the skeletons found in the Nemegt Basin are *Tarbosaurus*. Footprints found in the same geological formation suggest that tarbosaurs made up only about 5 per cent of the animals in the fauna.

'I have often wondered why we find so many *Tarbosaurus* skeletons in Mongolia,' Currie says. 'Perhaps the fact that we don't find duckbill dinosaur skeletons or small dinosaur skeletons relates to the fact that the tarbosaurs completely dismembered them and ate them 100 per cent.'

For many years, tyrannosaurs were believed to have been solitary hunters or scavengers that simply roamed the ancient environment and ate whatever they found and could easily catch. Currie is convinced that is not the full story, and the evidence supports him. He believes they were active predators. To build this theory, he uses both a practical approach and scientific data from dinosaur finds in Alberta and the Gobi Desert.

'Fundamentally, if you are talking about packing tyrannosaurs, potentially including *Tyrannosaurus rex*, you are not talking about scavengers; you are talking about predators because a pack of large scavengers would not be able to find enough food,' Currie says. 'There is no such thing as a pure scavenger among land-bound animals anyway. A big *T. rex* weighs the same as a big elephant. If they were scavengers, it would be impossible for them to find enough food to exist. Modern animals like hyenas can't do it either.'

Louis Jacobs, the vertebrate fossil expert who joined Currie on the Gobi expedition and at the lab in Korea where the Gobi finds were examined, believes that tarbosaurs were opportunistic eaters, making them both scavengers and predators. 'If you look at essentially any carnivore, they'll take any meal of meat if they can get it,' Jacobs maintains. 'So they'll scavenge if there's something to scavenge and they'll kill if there's something that they can kill. Once there's blood in the air, then I

'You are not talking about scavengers; you are talking about predators.'

would expect that any *Tarbosaurus* would take advantage of any weakness for a meal, so that if more than one could get in on a kill, they would do it. Whether there was a strategy as in wolves, where they run things down, or as in a pride of lions, where they're all chasing together, I don't know whether the entire social bond was that close. But if you have a large number of tarbosaurs on the landscape and you have something to eat, it's pretty reasonable to assume that there would be interaction, and as soon as the attack happened, the tarbosaurs would gather. [It's] the level of sophistication I'm not sure of.'

To determine the level of sophistication in tarbosaurs, Currie criss-crossed the world examining every piece of ancient and modern evidence he could find. He started by assessing the physical structure of the tarbosaurs and concluded that the ancient beasts were capable of doing some amazing things because of the way they were built. They were bipedal animals that walked on their hind legs somewhat like humans, but they were balanced in a very different way. Instead of being completely upright, they leaned pretty far forward, which is why they needed a long tail to counterbalance the weight in the front part of the body. Their legs were capable of powering them at relatively fast speeds, and it was the proportions of those legs that really gave them an advantage.

Tarbosaurus, particularly the younger tarbosaurs, had legs that were better proportioned for speed than human legs. Their legs were closer proportionally to those of ostriches, with the lower parts of the legs more elongated so that they could propel themselves at much higher speeds than even the fastest humans can manage.

'What we do know is that because of those leg proportions in *Tarbosaurus*, it was faster than anything it might have wanted to chase,' Currie says. 'So it's faster than the big duckbill dinosaur, and it's faster than any of the sauropods. It's certainly faster than the ankylosaurs.'

No trackways for tarbosaurs have been found, so their speed must be estimated. To determine a range of speeds, Currie studied the ostrich because of its nearly identical leg proportions. One of the fastest animals on the planet, the ostrich can achieve speeds of 45 mph. Currie both observed the ostrich in the field and dissected an ostrich leg. By comparing the muscles in the ostrich leg to the attachment points on the tarbosaur bones, he was able to draw a close parallel between the speeds of a juvenile tarbosaur and an ostrich.

'Having done the dissection on a ostrich, I think we can be fairly confident in saying that young tyrannosaurs would have been moving in the same kind of speed class as an ostrich,' Currie says. 'But as they got bigger, their bodies were getting larger at a faster rate than their legs were getting long. Although their legs were longer as adults (and intuitively you would think these animals must be getting faster) than as juveniles, they were changing proportions. We can do computer analyses and factor the relative changes as the animals age and grow up to determine whether in fact they were going as fast or slower.'

For advanced speed testing, Currie relied on the work of John Hutchinson at the Royal Veterinary College in Hatfield, England. Hutchinson used motion capture techniques to create advanced computer models to measure speed. These used stride lengths and the pressure of the foot hitting the ground to calculate speed. The conclusion reached was that the juvenile *Tarbosaurus* were significantly faster than the adults.

This supported a key aspect of Currie's theory, specifically that the young were much faster than their parents. Speed would have been critical for pack hunting. This is common in modern animals, as well as in humans. Based on this information, Currie was able to conclude that it is likely that the young tarbosaurs had a specific role in hunting, as the ones who used their speed to cut off the prey and drive it back towards the larger adults for the kill. If individuals of different ages and sizes had different body proportions and speeds, they probably had different roles, and it is quite possible that a group of different aged dinosaurs may have worked together in specific ways.

'The *Tarbosaurus* doesn't have ostrich proportions in its legs as an adult, but as a juvenile, it's a lot closer,' Currie says. 'This meant that the juveniles were a lot faster, and so if the adults were too slow to hunt in gangs, the juveniles were more capable to pursue and overtake the prey, and to corral it or harass it.'

Larry Witmer, the dinosaur brain expert, points out that this hunting scenario would have required a somewhat sophisticated thought process on the part of tarbosaurs. 'The key thing about being a communal hunter, or acting as part of a pack, is that each animal has some kind of role that they're playing,' Witmer says. 'Not all animals necessarily are going to be the ones making the kill. Some animals may be driving animals towards the other animals that are making a kill. The key thing in being a social cooperative hunter is, in a sense, you're deferring when you're going to get your meal. Your role may be to actually get your meal later. You might not be the one making the kill, but you're doing it on the understanding that you will be fed as part of that group.'

This is similar to the way modern-day lion prides hunt. Currie believes that tyrannosaurs may have mirrored the behaviour of lions as a group comprised of adults, sub-adults, juveniles and babies when they moved across the ancient landscapes. They snapped at each other, just as modern lions do, they squabbled, and they reaffirmed bonds. He concludes that young and old stayed together for a reason: hunting prey.

'We know from the *Albertosaurus* bone bed that the juveniles were hanging out with the adults,' he says. 'For the survival of the group, they presumably had a function and made a contribution to the "family". How do you explain it otherwise? There's no reason for them to be there for mating or anything else. If these animals were living together but hunting in a solitary way, then obviously the juveniles would be going after different things than the adults. But then what is that advantage of staying with the adults? There is no advantage. Physically, the young individuals were not capable of running, hunting or killing in the same way as the larger adults. And when the big, structured herds of horned or duckbilled dinosaurs were passing through their territories, how were they going to eat if they attacked as a random, unstructured mob?'

Watching lions in the African savannah gave Currie a vision of how the massive tyrannosaurs may have hunted together. The lions have a very

'Watching lions in the African savannah gave Currie a vision.'

good sense of smell and excellent vision, and their stability of gaze allows them to focus on prey while they are running. They also have great agility, allowing them to change direction at a moment's notice. Most importantly, they cooperate.

During a hunt, two splinter groups of young, agile lions broke from the main pride and sprinted off around the outskirts of a wildebeest herd to drive them back towards the other lions, with the intention that these other lions would make the kill. Without this ability, they would not be able to catch anything. Currie envisions tarbosaurs doing the same as they circled a group of hadrosaurs or sauropods, and worked together in gangs to track and bring down prey.

'The analogy is that a certain amount of hunting is more instinctual than you think,' Currie says. 'Of course, it's not solely instinct. There is something that is based on experience and their ability to learn. So probably with tyrannosaurs, like lions, the young ones don't always have the capability to take down the prey. When young lions are going after wildebeest, water buffalo, or even elephants, they don't have the strength or the knowledge of how to do it, so they will sometimes chase the prey and try and round it up and chase it back towards the adults, especially the big males. The big males are slower, but they are fierce and powerful. A big tyrannosaur is *really* powerful. Even though it was faster than its prey, it still had to get within a certain range to be effective. It didn't do it much good if it has to run 800 metres (half a mile) before it gets within striking distance because it would be pooped by the time it gets to them; whereas the young guys are fast enough and agile enough. This shows cooperation.

'It's very funny,' he adds. 'I don't know why that kind of thing is controversial, because in modern animals everything from fish to insects cooperate. And yet for some reason, because it's dinosaurs and you can't actually see it happening, people won't accept it as a possibility.'

Detractors contend that there is not enough evidence to conclude that dinosaurs hunted cooperatively. Though Jack Horner believes dinosaurs were social, he does not accept that they were cooperative hunters. 'There is really no physical evidence for it,' Horner says. 'In order to say something like that about a tyrannosaur, you would have to find something like that for a tyrannosaur. The fact is that you would have to be able to demonstrate taphonomically [from the study of the fossil record] that this is the best possible explanation.'

Despite the fact that the evidence is not conclusive, Currie's life work has led him to believe that it is in fact the best possible explanation. 'We know there is not enough evidence,' Currie acknowledges. 'The only evidence we could have that would be convincing is a time machine. That's the way it is. The reality is that if you don't put these ideas out there and they are not controversial, then they never get tested. People would never look for other lines of evidence. Being controversial is part of the process. It's a good, healthy thing. But still, sometimes I wonder why some people get so tied in a knot about it, because it is not a big deal. Modern animals do the full range of things that are suggested for dinosaurs. Animals of all evolutionary levels just need the right circumstances to develop certain behavioural responses. And clearly with dinosaurs, we had the right circumstances. We have animals that were hunting other animals that moved in huge herds, maybe thousands of animals. You can't move into a herd as one individual and expect to walk out with anything. Yes, you can follow the herd along and wait for something to die, but that's a tough way to make a living.'

Larry Witmer points out that just because a large dinosaur was a predator doesn't mean it would not have received some resistance from its prey – which

The Therizinosaurus *had a large hand claw that it could have used to defend itself against predators.*

is all the more reason that carnivorous dinosaurs would have hunted together.

'Being a predator is actually pretty risky business because the prey animal doesn't want to be eaten,' he says. 'In many cases, these animals have the tools

to fight back against the predator so [the predators] should be looking for means to reduce the risk to themselves. One way is to hunt as part of a group, to cooperate in how they're going to take down that prey. That reduces the risk for each individual predatory animal.'

Witmer points to a *Therizinosaurus* as an example. The *Therizinosaurus* had a large hand claw that it could have used to defend itself against predators. 'If we think about a *Tarbosaurus* trying to take down a *Therizinosaurus*, which is about the same size, on its own that would be a really risky proposition,' he says. 'But if tyrannosaurs like *Tarbosaurus* were acting cooperatively in a group, it potentially would have reduced its risk of being swiped by one of those metre-long claws from *Therizinosaurus*.'

For Currie, this is further evidence that the tarbosaur needed to hunt in gangs.

'There's no doubt at all that if tyrannosaurs like *Tarbosaurus* collected into packs, either temporarily for hunting or as more or less permanent structures, those animals would be able to take down much bigger prey than they would be capable of doing if they were hunting individually,' he says. 'There's good reason for *Tarbosaurus* to collect into packs if what they're doing is cooperatively hunting to take down the prey like sauropods.'

After spending time in the Gobi Desert and in Korea with Currie and studying his analysis, Yoshi Kobayashi, has also become increasingly more convinced that dinosaurs hunted cooperatively. 'Before we thought that those big carnivorous dinosaurs, including *Tarbosaurus*, were probably living as one, not as a pack,' he says. 'But it's getting common that many theropod dinosaurs may have lived as groups, including ostrich dinosaurs. We found this bone bed of ornithomimids that probably lived as a herd. That *Tarbosaurus* or *Tyrannosaurus* lived in packs makes more sense. Some people say that *Tyrannosaurus* is too big to run fast or chase prey. But the young *Tarbosaurus* (or *Tyrannosaurus*) could run fast – as fast as ostrich dinosaurs. They had that ability to chase and catch prey. Then maybe adults with the strong jaws and the big teeth can come and eat it. But they needed this strategy to live. I think that's a possibility that they lived in groups.'

The complete picture of the *Tarbosaurus* is building. This dinosaur was big, strong and fast. All of its anatomical features – from the massive jaws to its leg length, the shape of its tail and position of its eyes, nasal cavities and fore limbs – show this animal to be the perfect killer, designed to ambush, to scavenge, and ultimately to hunt. Its eyes face forward and afford it good vision; its massive tail provides balance; while its musculature shows it was faster and stronger than any other predator that existed in the Gobi during the Late Cretaceous. And now it also seems to have been more intelligent than

CGI of hunting tarbosaurs showing the older, larger animals hunting with the younger, smaller and faster animals. The faster animals corner prey, before the larger animals move in for the kill (bottom).

the animals it was hunting. Looking at the detailed anatomy of tarbosaurs, palaeontologists can learn a great deal about what this combination of brains and brawn may have been capable of doing.

Larry Witmer's CT scans allowed Currie to study the skull of a *Tarbosaurus* and to see inside the open area that contained the brain. In the brain case, Witmer and Currie were able to see the holes and re-create in a 3-D computerized model the wiring running through them. Because they know what that wiring looks like in the brain enclosure of tarbosaurs – even being very conservative with their analyses – they can compare what they are seeing to modern animals. This allows them to make comparisons between tarbosaurs and birds, tarbosaurs and crocodiles, and even tarbosaurs and mammals. Comparing the dinosaur's brain to that of modern animals allows the scientists to draw some conclusions about how well developed the different senses were in tarbosaurs.

'For example, we know that the olfactory bulb of the brain, which is right at the front, is connected with the sense of smell, and it's very, very well developed in all tyrannosaurs, including *Tarbosaurus*,' Currie says. 'We can look at the optic lobes which are connected with the optic region of the skull, and we can see that they're also well developed and that *Tarbosaurus* had a good sense of sight.

'We can look at the bones around the brain and we can see that there's all kinds of things going on

293 the dino gang theory comes to life

in the middle ear and the inner ear of *Tarbosaurus*,' he continues. 'First of all, these things allowed it to have a good sense of balance; but secondly, we can also see that there's some very sophisticated changes that have taken place compared to other meat-eating dinosaurs. *Tarbosaurus* and its relatives took air from the middle ear, the region behind the eardrums, through tubes that penetrated into the interior of bones surrounding the brain and ear. There they expanded into sinuses that not only lightened the skull, but allowed the animal to hear a greater range of sound wave frequencies. Secondly, it has connections between the right side and the left side of the skull, which allowed it to be able to hear what direction sounds were coming from. So that's also very sophisticated. All of these senses that we can see in *Tarbosaurus* suggest that probably this animal was capable of being a very good hunter. It would not need these improvements in sensory ability if it was a scavenger.'

When tyrannosaurs were first studied, palaeontologists were impressed by their size and apparent ferocity, but they did not give much consideration to them working together in gangs to hunt prey – a concept that makes them perhaps the most ferocious predators that ever lived. But we now know that the juvenile *Tarbosaurus* was fast like an ostrich, and that the adult *Tarbosaurus* was as ruthless and tenacious as a Komodo dragon, measured 12 metres (39 feet) long and stood 5 metres (16 feet) tall, tipped the scales at 5 tonnes, and was indeed very clever in comparison with its prey.

'When we first found tyrannosaurs, we were immediately struck by what giant predatory animals they were,' Witmer reflects. 'Looking back now, those old palaeontologists didn't really have a clue. The new studies going on right now are just revealing what a terrifying animal *T. rex* and other tyrannosaurs were. What we can see now is that *T. rex* was not that old, slow-moving, dull-witted predator of the past. Our new studies are showing that *T. rex* and other tyrannosaurs were fast, quick, agile animals that could potentially move together in communal packs. But more to the point, in their head they had all the tools to be effective predators – a sense of vision, hearing and smell to track prey, run them down and capture them. But even more importantly, locked up in their skull was a brain that gave them the capacity to hunt successfully and to kill. We immediately recognized that this was a terrifying animal, but that was just the start of it. Once we clothed those bones in flesh, we really start to see what this animal was about. It had all of the tools of a predator. We can drape those bones with muscles and skin and eyeballs and brains and really start to see the tools that this animal had to be an active, agile pursuit predator.'

Cooperative hunting implies an awareness of self and an understanding of the intentions of the others around you. It implies planning, strategy and thought of potential outcome. So *if* tyrannosaurs had this ability, it would mean that social intelligence existed millions of years before previously thought. And *if* they did, the implications could extend beyond dinosaur science and change entirely the way we think about the evolution of thought.

Currie, ever the cautious scientist, is wary of stretching the evidence, contemplating these big *if*s. 'All these things are certainly happening, but we can't even agree on them with modern animals, so to agree on them with dinosaurs is much more difficult. I would say that it is a little bit more controversial in that sense. There are people who still don't believe dogs think. What do you do? It's more of an argument about how you define these things rather than whether or not animals actually did (or do) them. If you set out and define what you mean by "think" and "plan" and get everybody to agree on it, then you would probably be okay. But the bottom line is that people believe (or they don't believe) that it's all instinct. And yet there are a lot of animals that learn things as they grow.'

Currie believes it's possible that dinosaurs had the ability to learn as they grew. 'In all probability, dinosaurs are in the same situation,' he says. 'When they were born, they probably didn't have the ability to do a lot of these things. By hanging around with their parents they were able to develop these skills in combination with instincts that they inherited. There are dogs that are instinctively good at retrieving things. There are other dogs that are instinctively good at swimming. But they still need training; it's always a combination of the two things – one, the genetics you are being controlled by, and two, the behaviour you develop from your experiences. Of course, we have no idea what capability dinosaurs had for learning. It's not necessarily associated with the size of the brain, but certainly in some animals it probably is. We can look at the brain and say that the main areas of a dog brain, which sccm to allow it to do morc

'It is not inconceivable that some dinosaurs did hunt cooperatively.'

complex things, are well developed in tyrannosaurs and therefore tyrannosaurs might have been able to do these things. But without a time machine, you can't go back and test it. Even with modern animals, in most cases it just hasn't been tested.'

Witmer analyses the gregarious hunting theory. 'The idea that tyrannosaurs hunted in packs is crazy for a lot of people,' he says. 'It's very controversial. We had always thought that tyrannosaurs hunted as solitary animals. So the idea that these animals hunted in packs has been viewed with some deal of scepticism. We scientists really want to see what the evidence is. This new idea is hitting us like a lighting bolt. On the other hand, what it opens up is whole new avenues for research. A lot of the things that we never looked for in terms of the biology of these animals we can now start to look for. The tyrannosaurs communal hunting – the gang theory of *Tyrannosaurus* – really opens up whole new avenues of study, things we can look at not just about *Tyrannosaurus* but about predatory dinosaurs in general, and it causes us to re-evaluate what we thought we knew about dinosaurs.'

Is Currie's fleshed out dino gangs theory and its implications controversial? 'Yes, of course,' he admits. 'You can always say maybe it doesn't hunt cooperatively, but that doesn't mean that it didn't happen that way. It's a possibility and you can't discount it. All you can say is, given the fact that modern animals from insects to mammals do it this way (young and fast animals living with the older, heavier, slower adults), it is not inconceivable that some dinosaurs did hunt cooperatively.'

Predatory gangs, pack mentality and cooperative hunting appear to be nothing new; Phil Currie has put together the evidence that it's very likely more than 70 million years old ...

A vast marshland tucked between mountain ranges hundreds of kilometres apart is alive with noise and activity. Flying reptiles and primitive birds with massive beaks and spindly wings swoop through the warm, steamy sky. Lizards of varying shapes, sizes and colours slither across the landscape. Turtles and crocodiles roam the banks of the rivers. In the centre of all this activity is a group of dinosaurs milling around and eating the vegetation.

Compared to the size of other inhabits of this ancient area, the dinosaurs are massive. These dinosaurs, which would later be named *Therizinosaurus*, are each more than 7 metres (23 feet) long and 3 metres (10 feet) tall and weigh some 3 tonnes. They have long necks, small skulls not much larger than the thickness of their necks, and relatively short tails. They stand on two legs and have bulging leg muscles, and their two, 2.5-metre (8-foot) long arms each have a trio of metre-long scythe-like claws on the end. With these claws, they flail at the trees and plants, foraging for their morning meal.

About a kilometre away out of the sight of the therizinosaurs, a pack of even larger dinosaurs is rousing itself for a feed. The adult creatures, which would one day become known as *Tarbosaurus*, measure nearly 12 metres (40 feet) in length, stand nearly 5 metres (16 feet) tall, and weigh in at 6 tonnes. The *Tarbosaurs* have two tiny, two-fingered forelimbs, and their long, heavy tails serve as counterweights to their heads and torsos, which are positioned in front of their hips. Every step they take shakes the ground in the immediate vicinity, sending the nearby wildlife scurrying for cover. The beast's skull is pure intimidation. At over a metre long, it has enormous eye sockets, and its mouth is packed with 64 banana-shaped teeth that are big enough to crush huge bones but also have sharp serrations on ridges along the front and back. This is the ultimate predator and the clear king of the landscape.

Two of the juvenile *Tarbosaurus*, which are similarly proportioned to the adults but about half the overall length, are biting and playing with each other. They circle one another in a mock hunt, butting their heads against each other's bodies. As one chases the other, their activities take them away from the pack into a clearing. Standing in front of them is a stray *Therizinosaurus* using its claws to rip a nest of terminates out of a log.

Hunting in gangs allows larger prey to be killed.

Instinctively, the young tarbosaurs stalk the much larger *Therizinosaurus*. Their pupils dilate

as they contemplate attacking. But as the young tarbosaurs close in, the *Therizinosaurus* fends them off by swinging its arms wildly. Slash! The metre-long claw from its right arm grazes one of the tarbosaurs. Blood sprays from the wound and the *Tarbosaurus* is spun around and thrown off balance. Quickly, the other *Tarbosaurus* hovers over its wounded playmate to prevent any further harm and lets out a plaintive roar that signals a matriarch *Tarbosaurus* that there is trouble. Even though it was two-against-one, the *Therizinosaurus* is able to escape and rejoin its herd.

The ostrich like gate of tarbosaurs is very effective.

The older, larger animals can attack much larger prey, if the prey has been prevented from escaping.

Not far away, a pack of a dozen dinosaurs with duck-like heads traipses along a nearby riverbed. These dinosaurs, which would later be classified as hadrosaurids and nicknamed 'duckbills', walk hunched over with their hips higher than their heads and on all fours, using their arms as a second set of legs. With their beak-like mouths, these dinosaurs – measuring up to 9 metres (30 feet) in length and weighing as much as 4 tonnes – bite the leaves off the horsetails and conifers close to the ground. After they have finished decimating the vegetation, the herd moves on in unison and with every stomp of the foot, the ground shakes violently.

The gang of tarbosaurs perks up at the sound of the shaking ground and starts moving in the direction of the noise. They begin slowly and then pick up speed, terrorizing everything in their path. Land animals scurry for shelter and flying reptiles spread their wings and squawk in an effort to protect their territory. But the tarbosaurs ignore them and focus on the bigger game: the herd of duckbills.

The hunt is on!

Tracking the prey by the sound of their pounding feet, the tarbosaurs plough through a huge cluster of trees and spot the duckbills. Sensing danger, the duckbills lift their forelimbs so they are standing on their legs and attempt to escape the threatening dino gang. The pack of tarbosaurs spreads out. Each *Tarbosaurus* seems to know its role. Two of younger tarbosaurs move to the right of the duckbill herd while two move to the left. Simultaneously, they crouch slightly, rise up onto their toes, and then run after the duckbills. Sprinting alongside the herd of prey, the young tarbosaurs keep their gaze firmly fixed on the lumbering duckbills.

Within seconds, the cunning and agile tarbosaurs have passed the running herd of duckbills and created a wall in front of them. The startled duckbills try to scatter, but as they reverse ground in an attempt to run for safety, they are driven back towards a gang of bloodthirsty adult tarbosaurs. As the massive predators lower their heads and prepare to strike their prey, a few of the duckbills are able to take advantage of the momentary distraction and escape. Others, however, are not so lucky.

One duckbill makes a poor decision to fight and is rammed in its flank by the head of the most ferocious tarbosaur. The duckbill stumbles, and the other tarbosaurs begin jerking their heads forward

'The startled duckbills are driven back towards a gang of bloodthirsty adult tarbosaurs.'

and snapping their jaws – killing machines sizing up their prey. As another duckbill tries to come to the rescue of its fallen brother, the tarbosaurs attack the two duckbills. The melee allows the other duckbills to flee.

In one attack, the *Tarbosaurus'* first bite rips the flesh of the duckbill's leg, and as the teeth penetrate the leathery skin, the serrated edges on the backs of the teeth cut the herbivore's tendons. The duckbill wails and crumples to the ground. The next bite from the carnivore yields a horrific crunch as the duckbill's vertebrae shatter. The second attack is a virtual replay of the first.

Now that the killer blows have been delivered, several other tarbosaurs surround the fallen duckbills and begin to feed. While the largest tarbosaurs maul the carcasses, the younger ones that corralled the herd wait their turn. Each

Tarbosaurus has several mouthfuls and then backs away so the others can eat. As the feeding frenzy winds down, every last bit of the duckbills – from toes to teeth – is eaten. When the tarbosaurs have finished, there is nothing left of the 2.5-tonne duckbill dinosaurs.

Their hunger sated, the larger tarbosaurs move into the shade. They spread apart, allowing each other ample space, and relax until their next meal. The marshland's sound track shifts from that of ferocious apex predators attacking their prey to animal life frolicking in the lush river delta system. An uneasy calm returns – until the next dino gang hunt.

epilogue

dinosaurs live on

In today's multimedia world, navigating the line between old school science and popular culture can be tricky. Phil Currie, for one, favours his theories and finds being available and understood by everyone, from school-age children to college students and from adults who grew up with an oversimplified image of dinosaurs to his learned fellow palaeontologists.

The cornerstone of palaeontologists' investigations is peer-reviewed scientific papers. Currie has authored numerous papers over the years, and he often works with other palaeontologists on them so they can pool their expertise and resources. In 2010, he edited a collection of scientific papers on the *Albertosaurus* bone bed for the *Canadian Journal of Earth Sciences* to mark the 100th anniversary of Brown's discovery – the bone bed that through study has provided some very convincing evidence of gregarious behaviour in carnivorous dinosaurs. He has several other papers in the works, including one on the tarbosaur sites in the Gobi Desert. But because peer-reviewed papers often take years to get published, Currie actively discusses his work in other media forums, and he wants to be sure that everyone with any interest in dinosaurs can understand his finds.

To reach a mainstream audience with their findings, Currie and the other scientists who joined him in the Gobi Desert and the labs in Korea and the US allowed their work to be filmed by London-based Atlantic Productions for a documentary that aired on the Discovery Channel in America. Atlantic has produced several landmark documentaries for the BBC, Discovery Channel, PBS and the History Channel, among other networks. Filming was done in secrecy under the working title 'Project P', to prevent information from being leaked before it was fully studied and debated by Currie and his colleagues.

Currie chose Atlantic for the quality of its films and the company's handling of the discovery of 'Ida', a 47-million-year-old primate fossil, whose unveiling in 2009 set off a worldwide media frenzy and a scientific debate over the possibility that it was a link between higher primates, such as monkeys and humans, and their more distant relatives, such as lemurs. He was introduced to Atlantic's founder, Anthony Geffen, through his close friend, palaeontologist Jorn Hurum, who was responsible for finding and studying Ida. He was particularly impressed with the way Atlantic's films have made history accessible and exciting to the average viewer, and hoped that the *Dino Gangs* film would bring attention to his work from professionals and possibly even help interest young people to pursue dinosaur palaeontology.

'The film will bring more focus on the *Albertosaurus* bone bed and I hope similar sites around the world,' Currie says. 'Even though you think that scientists only read scientific papers, the bottom line is that most scientists don't read them until they have to read them for some reason of their own. They are more likely to pick up on the [information] about it by watching or hearing about the film. Ultimately, what you hope to do is find new recruits in terms of getting kids interested in science.'

Over the years, Currie has seen an increase in college students taking an interest in dinosaur palaeontology, particularly among women. 'It's kind of funny that there are as many girls interested in dinosaurs as boys now,' says Currie, whose 2009–2010 ratio of male-to-female graduate students at the University of Alberta was 50:50. 'I think that it has a lot to do with the fact that the studies of dinosaurs have changed so dramatically. It's not just a big macho thing of digging up big game from ancient times.'

Currie is not shy about making bold pronouncements because he knows the only way to advance the scientific study of a theory is to present that theory. 'You hope to bring out other views, whether they are opposing or in your favour,' he says. 'It doesn't matter. The point is that other views and other people looking at the problem are going to look at more specimens and bring other angles to the debate. It never hurts. It was the controversy with the warm-blooded dinosaurs in the 1970s that helped spark the dinosaur renaissance. Even though they couldn't come up with a definite answer, a lot of people started looking at the idea, and now we feel pretty confident about some dinosaurs being warm-blooded. But it has taken three decades of research. All the controversy about dinosaur extinction? Same

'Currie's work completely changes the picture of how the planet's most dominate predators lived.'

thing. Now there are people actually doing fieldwork and legitimate research on it. The controversy started 100 years ago, and it's still raging.'

However, Currie is cautious to support his conclusions with scientific evidence and research. He hopes that the gaps in the historical and scientific record needed to prove his dino gangs theory will be filled over time, either by himself or by other palaeontologists who take his information and put it to the test.

'In spite of the fact that I believe that controversy is good, I also recognize the fact that sometimes things are said that shouldn't be said, or at least should be more diplomatically said,' Currie says. 'In the long run, however, you have to look at what happens. And I strongly believe that my theories about packing dinosaurs will, in the long run, be accepted by a lot of people because we're going to get more and more evidence to support it.'

Popular beliefs about dinosaurs have been largely shaped by what has been depicted in books and movies. The fierce carnivores have, by and large, been portrayed as solitary hunters. But with Currie's work, we can now conceive of the idea that these thrilling ancient beasts acted cooperatively. This completely changes the picture of how the planet's most dominate predators lived and it raises the possibility that social, strategic hunting started with dinosaurs. If this is true, what are the implications? If we have progressed to a place where cognitive behaviour is at its peak today, did it actually start 70 million years ago? What if these dinosaurs behaved in a similar way to modern predators, and to even the predators we know best – humans?

Palaeontologists may never be able to make these definitive conclusions. What they are willing to do is imagine how the ancient world functioned so we can relate this to ourselves. Because there may never be answers to these big questions, the unveiling of Currie's revolutionary finds will at least provide for more debate and study on a long journey to a fuller understanding of the world of dinosaurs.

'The rule of thumb when you're a palaeontologist is you say you don't know, because the fact is that as much as you do know, you still only know a small percentage of what you could know,' Currie says. 'So every time we answer one question we end up developing a dozen more that we may never have any chance at all of answering, simply because the fossil record is pretty spotty. We doubt that we've found and named more than 5 per cent of all dinosaurs that lived in the Late Cretaceous and that is one reason why we keep finding new dinosaur species all the time.'

Currie has spent a lifetime searching for answers about the dinosaurs' world. From the days he dug through cereal boxes looking for a plastic *T. rex* to his remarkable rediscovery of the *Albertosaurus* bone bed, and from his days as a child spending afternoons in the Royal Ontario Museum in Toronto to his expeditions in the hostile Gobi Desert, his fascination with the lives of dinosaurs has consumed him. As his journey to reach a more complete understanding of these ancient animals continues, he has also come to see a connection between their existence and our own.

'One reason people are so fascinated with dinosaurs is that it makes us think about our own future,' Currie says. 'If we as humans have been around for only 4 or 5 million years, consider that dinosaurs dominated the world for 150 million years. In 20 million years, are we still going to be around? Probably if you took a poll on it, most people would say we don't have much of a chance the way we are going. The lesson I feel is important now is that it's not just about dinosaurs or humans – it's every living thing. Right now, we are wiping out biological diversity at an unprecedented rate. Given the fact that the loss of dinosaurian diversity was one of things leading up to their extinction, we should think twice about further reducing the diversity of modern species, even plants. It's a dangerous thing to do. We are removing our options for long-term survival slowly but surely. We may not find the answer to preventing our extinction by studying dinosaurs, but the reasons for their extinction after being successful for so long are worthy of contemplation.'

index

Entries in *italics* indicate photographs and illustrations.

acknowledgements

This book resembled a dinosaur dig with many people contributing their strengths to the project and then chipping away until the final product came together.

First and foremost, Phil Currie enthusiastically gave his time and knowledge to the project. The world of dinosaurs is one he knows intimately, and his ability to communicate highly technical aspects in a digestible form sets him apart from those in his trade. His wife, Eva Koppelhus, carved the time out of Phil's schedule and added a few field stories of her own.

A group of world class scientists, Yuong Nam Lee, Yoshi Kobayashi, David Eberth, Larry Witmer, Donald Henderson, Louis Jacobs, and John Hutchinson, provided valuable insights into specialized areas of dinosaur palaeontology. Jack Horner, who was not part of the documentary film for which this book is the companion, also provided his insights into dinosaur behavior.

The team at HarperCollins was led by Myles Archibald, who oversaw all aspects of the book.

The team at Atlantic Productions made the book and the film happen. Anthony Geffen, a producer of the highest order, brought everyone who worked on the project together, and James Taylor served as the